Making the Military Moral provides a rich tapestry of diverse issues in military ethics and military ethics education. One of its great strengths is the diversity of perspectives, both disciplinary and international, it offers on these issues. This volume is an important contribution to an increasingly global conversation on issue of military ethics.

Martin L. Cook, *United States Naval War College*

Making the Military Moral

This book offers a critical analysis, both theoretical and practical, of ethics education in the military.

In the twenty-first century, it has become increasingly important to ensure that the armed forces of Western and other democracies fight justly and behave ethically. The 'good soldier' has to be not only professionally skilled but morally intelligent. At a time of relentless media scrutiny, the publicising of incidents of morally and legally unacceptable behaviour, such as the gross mistreatment of prisoners and the torture of suspected terrorists, can do much to undermine the credibility of those who claim to hold the moral high ground in any particular conflict. Written by an international team of academic theorists and military practitioners, this volume provides inter-disciplinary insights into the present state, and the future, of ethics education in the militaries of Western democracies. The contributors critically address the central question of whether such education is sufficient to prepare members of the armed forces to face the peculiar challenges of conflict environments that are now primarily 'wars among the people', in which the opposing combatants may have little or no regard for human life and fail to discriminate between soldiers and civilians when choosing their targets. Drawing lessons from recent examples of unethical conduct, this original book offers insightful and constructive advice, both theoretical and practical, as to how situations can be improved and on the means that could and should be employed towards this end.

This book will be of much interest to students of military studies, ethics and international relations.

Don Carrick is an Honorary Research Fellow in Applied Ethics at the University of Leeds, UK.

James Connelly is Professor of Political Theory at the University of Hull, UK, where he is Director of the Institute of Applied Ethics.

David Whetham is Reader in Military Ethics in the Defence Studies Department of King's College London, UK.

Military and Defence Ethics

Series Editors: Don Carrick
University of Leeds, UK
James Connelly
University of Hull, UK
George Lucas
Naval Postgraduate School, USA
Paul Robinson
University of Ottawa, Canada

There is an urgent and growing need for all those involved in matters of national defence – from policy makers to armaments manufacturers to members of the armed forces – to behave, and to be seen to behave, ethically. The ethical dimensions of making decisions and taking action in the defence arena are the subject of intense and ongoing media interest and public scrutiny. it is vital that all those involved be given the benefit of the finest possible advice and support. such advice is best sought from those who have great practical experience or theoretical wisdom (or both) in their particular field and publication of their work in this series will ensure that it is readily accessible to all who need it.

In the series:

The Warrior, Military Ethics and Contemporary Warfare
Achilles Goes Asymmetrical
Pauline M. Kaurin

When Soldiers Say No
Selective Conscientious Objection in the Modern Military
Edited by Andrea Ellner, Paul Robinson and David Whetham

From Northern Ireland to Afghanistan
British Military Intelligence Operations, Ethics and Human Rights
Jon Moran

Making the Military Moral
Contemporary Challenges and Responses in Military Ethics Education
Edited by Don Carrick, James Connelly and David Whetham

Making the Military Moral

Contemporary Challenges and Responses in Military Ethics Education

Edited by Don Carrick, James Connelly and David Whetham

LONDON AND NEW YORK

First published 2018 by Routledge

2 Park Square, Milton Park, Abingdon, Oxfordshire OX14 4RN

52 Vanderbilt Avenue, New York, NY 10017

Routledge is an imprint of the Taylor & Francis Group, an informa business

First issued in paperback 2020

British Library Cataloguing in Publication Data
A catalogue record for this book is available from the British Library

Library of Congress Cataloging in Publication Data
Names: Carrick, Don, editor of compilation. | Connelly, James, editor of compilation. | Whetham, David, editor of compilation.
Title: Making the military moral : contemporary challenges and responses in military ethics education / edited by Don Carrick, James Connelly and David Whetham.
Other titles: Contemporary challenges and responses in military ethics education
Description: Milton Park, Abingdon, Oxon ; New York, NY : Routledge, [2018] |
Series: Military and defence ethics | Includes bibliographical references and index.
Identifiers: LCCN 2017015246| ISBN 9781472412058 (hardback) | ISBN 9781315593388 (ebook)
Subjects: LCSH: Military ethics–Study and teaching. | War–Moral and ethical aspects. | Military education.
Classification: LCC U22 .M319 2018 | DDC 172/.42–dc23
LC record available at https://lccn.loc.gov/2017015246

ISBN: 978-1-4724-1205-8 (hbk)
ISBN: 978-0-367-66758-0 (pbk)

Typeset in Times New Roman
by Wearset Ltd. Boldon, Tyne and Wear

Contents

Contributors

Janne Aalto is Senior Chaplain at the Defence Command Finland. He has served on several units in Finnish Defence Forces. He holds a Doctor of Military Science degree in Military Pedagogy from the National Defence University Finland and a Master of Theology degree from Helsinki University.

Stéphanie A.H. Bélanger is the Interim Co-Scientific Director of the Canadian Institute for Military and Veteran Health Research and the Co-Editor in Chief of the Journal of Military, Veteran and Family Health. She is Professor at the French Department of the Royal Military College of Canada.

Peter Bradley retired from the Canadian Armed Forces as a Lieutenant-Colonel in 2004. At the time of his retirement he was Head of the Military Psychology and Leadership Department at the Royal Military College of Canada. During his academic career he was the author of numerous articles on ethical decision making in the military.

Don Carrick is an Honorary Research Fellow in applied ethics at the University of Leeds. He was one of the founders of the Military Ethics Education Network (the MEEN) and has latterly served with James Connelly as the MEEN's co-director. He publishes in military medical ethics and the Just War Theory.

James Connelly is Professor of Political Theory and Director of the Institute of Applied Ethics at the University of Hull. He was one of the founder members of the Military Ethics Education Network (the MEEN). He publishes in ethics, political theory and philosophy.

David Fisher had a long and distinguished career in the UK Civil Service, serving as a senior official in the Ministry of Defence, the Foreign Office, and the Cabinet Office. He published widely on war and morality and the just war tradition, and latterly became a much respected member of the War Studies Department at King's College London, where he held a Teaching Fellowship.

George R. Lucas Jr is Professor Emeritus at the United States Naval Academy. He is the author of 'Ethics and Cyber Warfare' and 'Military Ethics: What Everyone Needs to Know' both published by Oxford University Press (2016).

Allister MacIntyre is a Psychology Professor, and former Department Head, at the Royal Military College of Canada (RMCC). He served 31 years with the Canadian Forces and he has been an executive member of the Canadian Psychological Association since 2002.

Michelle Moore, MA, is a Lieutenant Navy in the Canadian Armed Forces. She is currently serving as a Workplace Relations Coordinator, facilitating formation level ethics training, harassment prevention, and personnel-related investigations. She has a Masters in War Studies from the Royal Military College of Canada.

Sally Rohan is a Senior Lecturer in the Defence Studies Department of Kings College London, based at the UK's Joint Services Command and Staff College. Her research focuses on professional military education and military ethics.

John Thomas served in the Royal Air Force for 32 years, retiring as an Air Commodore. His responsibilities have included serving as a Director in the UK's Joint Services Command and Staff College, and as a Deputy Director of the Air Staff in the UK MOD. He is currently co-Executive Director of the International Society for Military Ethics in Europe (Euro-ISME).

David Whetham is Reader in Military Ethics in the Defence Studies Department of King's College London, based at the UK's Joint Services Command and Staff College. He has written widely on the subject of military ethics and is a founding member and Vice President of Euro ISME as well as being the Director of the King's Centre for Military Ethics.

George R. Wilkes is the Director of the Project on Religion and Ethics in the Making of War and Peace at The University of Edinburgh. He was previously a Fellow at St Edmunds College, Cambridge. His research currently focuses on effectiveness and cultural, religious or ideological diversity in the delivery of military ethics education.

Acknowledgements

We owe a large debt of gratitude to very many people for their help in bringing this volume to publication, but mainly to our loyal team of authors who, with exemplary patience, tolerated the exceptionally long period of gestation that occurred before they saw their splendid efforts come to fruition and publication. To them, we extend our most grateful thanks.

We also owe a debt to the Leverhulme Trust who funded the project 'International Network for the study of ethics education in the military' (commonly known as the Military Ethics Education Network). This award facilitated the travel and research which made possible the creation of both this book and also the prior books in what was then Ashgate Publishing's Military and Defence Ethics series.

Finally, we owe a particular debt of gratitude to the late David Fisher, whose untimely death in 2014 deprived us all of one of the leading academics and practitioners in the world of military ethics. This volume therefore constitutes a fitting memorial to the man who not only contributed a fine chapter to the book, but also suggested the title that it now bears: this work is therefore dedicated to David, with respect and affection.

The Editors

1   On making the military moral[1]

James Connelly

Introduction

It is customary with edited collections of essays to write an introduction summarising those essays. This chapter introduces the book more obliquely, through reflection on the idea of military ethics and military ethics education, in particular.

Some people, when they hear the phrase 'military ethics' respond by remarking that, like 'business ethics' or 'legal ethics', it is a contradiction in terms. It is a rather depressing question, frequently unworthy of its author. But we shouldn't leave the matter there. Of course, it is a thoughtless and shallow remark, but it has genuine roots. It reveals a certain idealism, the sort of idealism which is the obverse side of cynicism: in an ideal world, military ethics would be redundant because military force would be unnecessary; but in the world as it is, military action is both necessary and thoroughly evil. Either way, military ethics, it is averred, is a contradiction in terms. However, the more interesting point is that the response trades on an ambivalence in the term 'ethics' itself. On the one hand, we have ethics as a system of moral judgements delineating 'moral' or 'immoral' actions, and, on the other, we have ethics as the analysis and critical reflection on conduct. This ambivalence is inherent in the tensions which arise in teaching ethics to the military.

Is there, as Alasdair MacIntyre avers, a crisis in military ethics? (MacIntyre 2015). And is there a crisis in the teaching of military ethics, the imparting of ethical reasoning and principles of members of the military? My argument is that there is no crisis, but there is a rapid evolution both in teaching and in what is to be taught and to whom. In the military, we have moved decisively away from old practices of moral deference to a new practice of critical moral quasi-autonomy in which those who are led are co-creators in the projects in which their leaders exercise authority.

To begin. We can distinguish ethics as moral philosophy (in the form of meta-ethics or normative ethics), ethics as a set of ethical principles, and military ethics, which can itself be a meta-discipline, reflecting on the conditions of ethics, or a normative set of principles designed to apply to action. Military ethics is a branch of applied ethics which comprises elements of meta-ethics and normative ethics; successfully teaching ethics to the military has to reflect this.

With this in mind, let us question the phrase 'making the military moral'. *Prima facie* this seems an impossible task. The underlying question is whether it is possible to *make anyone* moral; a related question is what sort of collective entity is it to which we apply the appellation 'military'? Is it a genuine collective with an agency as well as an ethos, or simply a collection of individuals whose individual morality we can shape and mould, directly or indirectly?

All applied ethics is hybrid; military ethics is perhaps more hybrid than any other form of applied ethics, if only because it covers such a wide range of possible things, as George Lucas points out in his introduction to the *Routledge Companion to Military Ethics* (2015). 'Making the military moral' covers, for example, making members of the military moral, making the military as an institution or collective entity moral, and making what it does moral. The word 'moral' here is used more or less as a synonym for 'good'; that is, it is not used in the broader sense of a predicate which can be attached to any action, including military action. But stated thus, 'making the military good' sounds insufferably moralistic; I am sure there are those who would say that 'making the military moral' sounds only a little better. Can there be military ethics? Yes – because there can be an ethics, critical reflection upon, for anything to which moral predicates can be attached. Can the military be made morally good? That is a more open question, which this book seeks to examine. Let us proceed by asking how we teach military ethics and let some of the answers to the issues raised above unfold as we do so.

Methods in military ethics I: preparing the ground

My argument is that military ethics, as part of the military curriculum, should primarily be devoted to inculcating the ability to engage in moral reasoning. In this statement, I am implicitly distinguishing military ethics from the acquiring of military virtues, which are developed in practice and not in the classroom, although they can be reflected upon in the classroom. Military ethics, that is, does not seek to make students into moral philosophers, neither is it concerned with rote learning and regurgitation of abstract moral theories per se. But neither is it anti-theoretical, and it should certainly not be merely a matter of moralising, moral exhortation, or injunctions to develop one's character. And it most certainly should not fall into an intuitionist trap where it is assumed without argument that 'we know the right thing to do' or possess a 'moral compass' which will act as a sure guide to understanding what is morally required in a situation in which we find ourselves. It is not that all talk of intuition or moral feeling is inappropriate, but it is and can be at best only a starting point: both the springs of ethical norms and the practice of ethical reasoning require work. And we might remind ourselves that compasses have to be designed and constructed painstakingly in such a way as to reliably point north. Military ethics education presupposes not that we already possess a properly working moral compass but that we need to construct a moral compass for use on specifically military terrain.[2]

In approaching these issues, ethics courses often begin with a ground-clearing exercise. The form this takes is highly dependent on country and culture. In some places it might require a demolition of the idea that moral intuition, the presumed possession of an infallible moral compass, is all that is required in reasoning and action; in others, it might require critical assessment of the relationship between moral reasoning and religious authority or divine command. I have witnessed introductory discussion, for example, of the dilemma presented in Plato's *Euthyphro*. In this dilemma, it is accepted both that the Gods command what is good, but also that they might have commanded otherwise. What follows? Is 'the good' good because *God commands it*? or does God command it *because it is good*? In the latter, there must exist another independent standard of judgement by which to judge that what the Gods say is good is indeed good. The reason for teaching the *Euthyphro* in a country where naïve religious views of ethics dominate is precisely to enable students to see the point and purpose of ethics education and moral reasoning because otherwise they are likely to believe that it can have no value or claim on them.

It is important to note that this raises the question of who is qualified to teach ethics and in what sense there can be expertise in ethics. Here we need to distinguish acting morally from the ability to analyse the conditions of moral conduct. The point is that by a moral expert we do not mean a moral hero: rather, we mean someone able to discuss moral theory in its application to military practice. In practice, we sometimes find a considerable gap in understanding and expectations between teachers who have a philosophical background and those who do not, and this sometimes leads to a failure to distinguish moral reasoning from moralising.

In the course of a research trip I heard a teacher of military ethics make what, to me, seemed to be an elementary mistake, which is to assume that moral predicates can be attached only to some actions and not to others. The question was: 'which sorts of things are moral choices and which are not?'; the answer was 'some things, such as what to have in your sandwich at lunchtime, are not moral choices, but others, such as whether to lie or kill, are moral choices'. Any environmentalist or vegetarian would immediately answer that what you have in your sandwich is a complex moral matter, where the issues to take account of include environmental degradation and climate change, the ability to feed the world's population, animal suffering and death, and so on.

Of course, it would be foolish to doubt that there is an issue of relative moral weight or moral considerability, to say that moral predicates attach only to some actions and not to others is a category confusion as well as being based on the false analogy of actions as natural kinds, as though they could be categorised as different types of metal or mineral. My view is that, on the contrary, it is intrinsic to anything we can call an action that moral predicates attach to it. We can ask, that is, whether it is right or wrong or good or bad, whether it should be performed or avoided, and so on. Judging relative significance and moral considerability is vital, of course, but itself requires an act of moral judgement and it presupposes that although nothing is exempt from moral consideration, successful moral reasoning lies precisely in judging the relative weight of different claims on our moral attention.

Methods in military ethics II: ethics requires judgement and cannot be reduced to law

Sometimes the need for reflection on ethics in the military is denied by those who think that the law and various codes and protocols cover everything: knowledge of these, it is claimed, is all that is needed, and moral reasoning as an independent activity is a redundant luxury. There is clearly a relationship between ethics and the law, but they are not identical, and this is for at least three reasons. The first is that all laws, rules and principles require judgement to apply them, and this cannot be provided by the law or rules and principles themselves; second, ethics is concrete where law is not – law is necessarily general; third, ethics includes critical assessment of all action, including legal action – the law is itself subject to moral evaluation and can be modified as a consequence of that evaluation. Hence addressing the moral dimension requires attention to appropriate forms of moral reasoning and these cannot be reduced to a ghostly ethical double of a legal code. Moral reasoning is inescapable: it cannot be reduced to laws, rules, tick boxes or codes; there are no algorithms which can be relied on to provide answers in the absence of moral reasoning; and also to give a sense of how moral reasoning in military ethics proceeds.

Methods in military ethics III: making it real? – The example of torture

Examples, 'what ifs', and case studies are a part of the standard armoury of the teacher in a military ethics class. But what is their purpose? Do they really help in developing moral reasoning? One of the difficulties is that, unless their purpose is thought through carefully, they are vulnerable to the charge of mere irrelevance. They can be easily hijacked by either the heroic exemplar stirring up the troops, or asking whether students would have the moral courage or tenacity to respond in the way that is deemed proper, no matter how hard it might be to do so. This might have some value, but it is surely of a very limited kind if the purpose of military ethics education is to enable students to engage in appropriate forms of moral reasoning.

Again, examples and cases studies are self-limiting if they are regarded as a simple matter of moral intuition: this is precisely because a good case study will recognize that there are moral uncertainties and ambiguities. This is the source of one of the apparent tensions in the way in which ethics education as a whole is dealt with. In particular, should they be taught with or without theory?

Examples and case studies can be divided into different categories: they do not all work in the same way. For example, Stephen Coleman of the Australian Defence College distinguishes two types of ethical problem: tests of integrity and moral dilemmas (Coleman 2013: 5). In a test of integrity, we know the right thing to do, but don't know whether we have the moral strength to be able to do it. Thus, it is a problem of character, not knowledge or reasoning: those within the military who focus on character training and development tend to focus on this to the exclusion of analysis of moral reasoning and moral dilemmas. It should be stated that moral integrity is clearly important and hence so is training people to have the appropriate character and virtues. But if one takes the view that what is required is the ability to reason morally in certain situations, how is this best done? A start would be by focussing on case studies that bring out ethical dilemmas. An ethical dilemma 'is a situation in which a person is faced with a number of choices, often a number of bad choices, and has to work out what is the right thing to do in that particular situation' (Coleman 2013: 5).

The point about examples which require students to reason their way into the complexities of moral dilemmas is precisely that they are aids to disciplined reflection and therefore have to have the appropriate character, that is, they have to bring out a number of salient issues illustrative of something wider than the example itself and demonstrating the nature of moral reasoning and its principles.

Do examples have to be realistic? The short answer is that if their purpose is to cast light on a particular moral concept then they do not; but if their purpose is to allow us to engage in reasoning through the complexities of a piece of concrete moral reasoning and honing our powers of moral judgement, they do. Realistic might mean real, or it might mean having all the relevant features of something that is real. Fantasy thinking in military ethics is no help to anyone. The teaching of ethics through example and case studies, if they are to satisfy these criteria and properly engage students in moral reasoning, requires teachers who approach them in the right manner. Earlier we suggested that exhortation and examples of moral courage might have their place; but their place is not in the analysis of case studies, and teachers have to avoid them. A teacher will probably have to be well trained in moral philosophy, but at the same time not be trying to make their students into moral philosophers, but into moral reasoners. Teachers with a theoretical background but attuned to practical exigencies and the responsibilities of command are therefore ideal, if they can be found. But if there is a choice, then the former must take precedence, because the latter will be inculcated both by experience and in the normal course of military training.

Let us consider a little more deeply the issues at stake here. Following Henry Shue (2009), we might fruitfully draw the distinction between all-things-considered judgements and judgements that follow from reasoning based upon abstraction and hypothetical cases in which the only considerations are mere conceivability, not the difficulties of practical judgement. This is especially important when considering exceptions to general rules. This has a high cost, the cost of having to immerse oneself in both the technical and practical detail of a case because to do otherwise is to run the risk of leaving out of account vital features of the problem.

To illustrate some of the points made above, let us take the example, increasingly (to some, distressingly) common, of scenarios in which torture is represented as being acceptable. There are many arguments to the effect that in some cases, (typically 'ticking bomb' cases) in which it is argued that where information can be acquired which will prevent a greater harm, it is justifiable to torture whoever possesses this knowledge. Outside the realms of fictional representation, and assuming for the sake of argument that the prohibition against torture is not absolute but conditional on its efficacy alone, could the conditions ever be obtained in which torture might be justified? How plausible is it to maintain that these conditions might genuinely arise in practice?

The case for torture in exceptional cases hinges on whether the scenario can be specified in sufficient detail to be plausible enough to sustain the conclusion its advocates seek. Many argue that it cannot. Why? First, we need to know that all things really have been considered. If things are to be discussed as if they were real-world cases we need to make sure that the 'real world' element is genuine. We cannot cherry pick which aspects of the world we accept and which we do not. It is clearly far easier to make a case for torture in an abstract or idealised case where there is no constraint on the degree of abstraction or idealisation we allow ourselves, than it is in a real-world case where we have to consider locating the practice in its historical and social setting. However, advocates of torture argue that in an all-things-considered case, where we are considering not merely the abstract morality of torture per se, but putative real-world cases in which torture might be (although repugnant) a plausible means to an end. They are thereby committed to spelling out a situation in which *prima facie* we are inclined to say 'here, under these special, these unique circumstances, torture might be deemed acceptable, although this does not mean that we regard it as *generally* acceptable'. Hence, all-things-considered, the scenario becomes a crucial test case. My contention, in which I agree with Bob Brecher (2007), is that at this point the all-things-considered view exposes not the strength but the weakness of the case for torture. Advocates of torture are adept at asking us, hypothetically, whether we would be in favour of torture in this or that situation, and in doing so they allow themselves imaginative free rein in developing scenarios in which we are inexorably led to agree. They insinuate that although (in the abstract) torture is unacceptable, wouldn't we in practice think otherwise if placed in a certain situation? The problem, however, is that they trade on the idea of the all-things-considered judgement whilst leaving crucial features unspecified and unaccounted for. They then write blank cheques in their own favour on the basis of unspecifiable, unrealistic or unacceptable assumptions. It seems to be that the more that the conditions which might conceivably justify torture in practice are specified, the more implausible the scenario becomes. The result is that torture in practice is revealed to be, on that occasion, both morally wrong, practically pointless, and also – this is crucial – as presupposing a general practice of torture in order to make its occasional practice plausible. Hence the justification of torture in practice is undermined by the requirement for the practice of torture. Why is this? Because torture is presented not as an ongoing institutionalised practice but as a one-off, exceptional and urgent requirement. The problem is that torture is not a job for amateurs – it requires expertise and practice. But if it really is the exception, a skilled torturer will not be available, and if a skilled torturer is ready at hand, this can only be because there is a general practice of torture, in which case it is not an exception.

To elaborate the points made above. We cannot assume perfect knowledge. We therefore cannot simply assume that we will know that the proposed victim of torture (whom we happen to have captured at just the right moment, not too soon and not too late) knows what we want to know and will break under torture. Again, we have to have good reason to believe that the torturer is capable of extracting information quickly from someone prepared to resist torture applied by an expert, rather than (in a hypothetical exceptional case) by someone who is not a trained torturer. And so on for all other aspects of the case. It appears to be that the more the details are spelt out, the more implausible the picture becomes. Indeed, it is a plausible scenario at all if the details are left relatively vague, but in that case it ceases to be an example of a serious practical judgement and turns into an exercise in moral fancy.

A general conclusion that follows from these considerations is that in the sort of concrete moral reasoning required in serious military ethics, we need to spell out the social institutions in which our actions are embedded. Many examples in moral philosophy are vitiated by a failure to recognise what we might term their vertical and horizontal setting, that is, how they came to be and how they are located within other situations and institutional arrangements. In the example above, we cannot assume that we have a competent torturer to hand if we have defined the case as an exception which excludes their presence. If, on the other hand, we design institutional arrangements into the example, it is no longer concerned as an exceptional case. We cannot have it both ways. Hence Shue remarks that:

> In the perfect case for torture, while torture is rare because restricted to such appropriate cases, the torture is perfectly successful: suddenly someone with no experience or training, who has never tortured anyone before, quickly extracts vital information from someone dedicated to withholding that very information. This is a sociological fantasy. We have abstracted from the social basis – the institutional context – necessary for the practice of torture. For torture is a practice. Practitioners who do not practice will not be very good at what they do.
>
> (Shue 2009: 314)

Methods in military ethics IV: abstracting, idealising and illustrating

Someone might reply that examples are necessarily hypothetical: but by this do we mean something that might happen, or merely, as noted above, a case that can be constructed to illustrate or bring out a moral point? And is the point of coming up with a plausible scenario to prove conclusively that one or another thing should (or should not be done) or to use it as a way of drawing attention to the relevant moral distinctions to be drawn in making a judgement? If the latter, do we need all the empirical apparatus or just enough to bring out the moral issue? A related point is that examples are often chosen not to illustrate a moral point, but to test a particular moral theory. In 'all-things-considered' judgements we really need either to be dealing with the full complexity of real cases or with examples which deal with the full, if hypothetical, complexity of real cases. In abstract, hypothetical cases, however, we might be focussing our attention solely on certain ethical ideas or principles. In the latter case, we can afford to consider examples which are less empirically rich in order to draw out those elements of moral reasoning we wish to get clear. In such a case we are not arguing to a practical decision but rather to a theoretical conclusion or a theoretical point.

To summarise, I would like to make a few general observations on the use of examples. First, they should be appropriate to the seriousness of the moral issue under consideration; trivial examples are inadequate if they fail to bring out the moral seriousness of cases to which their conclusions might be applied. Certainly, if we are concerned with all-things-considered judgements, it would seem absurd to choose slight examples. Second, should all examples be realistic? Obviously, in some cases at least, realism is important, as is the recognition of the distinction between abstraction and idealisation (O'Neill 1988). O'Neill argues that abstraction is necessary in all reasoning, but it can be contrasted with:

> an idealized account or theory [which] not merely omits certain predicates that are true of the matter to be considered but adds predicates that are false of the matter to be considered. Idealization requires abstraction, but they are not the same thing. The omission of some predicate, F, is not tantamount to the addition of the predicate, $-F$. 'Omission' of a predicate in abstracting merely means that nothing is allowed to rest on the predicate's being satisfied or not satisfied. By contrast, when we idealize, we add a predicate to a theory, and the theory applies only where that predicate is satisfied.
>
> (O'Neill 1988: 712)

This is an important point. It is easy to criticise any example for being abstract, but given that any example or case study will be abstract to some degree, this is a self-defeating point. The issue is whether the abstraction is materially misleading. In the case of idealisation, abstraction is materially misleading because we are bidden to take an ideal case as though it were a normal or real case and then to reason from what could only be true of the ideal to what is true in real life. Any example or case study which requires this is, at least, misleading and often simply fallacious. This should not, however, be assimilated to what might be termed outlandish cases which appear to be absurd, but which are designed not to persuade per se, but to bring out the salient features of an ethical theory or concept. Take Bernard Williams's well-known example of Jim and the Indians:

> Jim finds himself in the central square of a small South American town. Tied up against the wall are a row of twenty Indians ... in front of them

several armed men in uniform. ... [T]he captain in charge ... establishes that [Jim] got there by accident while on a botanical expedition, [and] explains that the Indians are a random group of the inhabitants who, after recent acts of protest against the government, are just about to be killed to remind other possible protestors of the advantages of not protesting. However, since Jim is an honoured visitor from another land, the captain is happy to offer him a guest's privilege of killing one of the Indians himself. If Jim accepts, then as a special mark of the occasion, the other Indians will be let off. Of course, if Jim refuses, then there is no special occasion, and Pedro here will do what he was about to do when Jim arrived, and kill them all. Jim ... wonders whether if he got hold of a gun, he could hold the captain, Pedro and the rest of the soldiers to threat, but it is quite clear ... [that] any attempt at that sort of thing will mean that all the Indians will be killed, and himself. The men against the wall, and the other villagers understand the situation, and are obviously begging him to accept. What should he do?

(Williams 1973, 98–99)

This is probably not a realistic example: but its purpose does not hinge on its realism, but rather on its ability to bring out the features of utilitarianism as a moral theory which judges actions by their consequences and not by the virtues of the agent, or his or her intentions. I submit that the example does not have to be realistic in the way that a fully developed case study in which one has to decide on a course of action has to be. It is not like the ticking bomb/torture scenario, because its dual function is to test our moral intuitions over various distinctions which it is held utilitarianism fails to account for and also to test the application of principles in different contexts.

By contrast, the ticking bomb case is designed not just to draw out issues related to our general moral reasoning, but to compel a particular conclusion. Again, the example of Jim and the Indians differs from the ticking bomb case because the epistemic conditions are less stringent. Jim did not require special knowledge or information and there is no problem in assuming that he knows all he needs to know to be faced with a serious and pressing moral dilemma. In the ticking bomb case participants are obliged to know a tremendous amount; and most of what they need to know it is either impossible for them to know under the stipulated conditions; or, if it is claimed that they know it, the example becomes self-refuting. Thus, an example such as Jim and the Indians, which looks outlandish, might nonetheless be a perfectly good example, but another, which appears to be plausible, is not.

Generally, in military ethics, we then need to distinguish not two but three types of example. Realistic ones which exemplify moral dilemmas; tests of integrity; and examples designed to test or bring out features of a moral concept or theory rather than to lead us to engage in actual or simulated moral judgement. Only the first can be fully articulated case studies, and it is these that are of most value in military ethics education. And we should not, when engaging in moral reasoning, simply bring along abstractions and try to make them fit.

On the contrary, the situation is primary and we should immerse ourselves in it, think our way through it and develop our understanding of its morally relevant features. Moral theories can guide us by telling us which kinds of features would be salient if they were present, but it is our responsibility to judge whether they are really present in this case or not.[3]

Methods in military ethics V: thinking in triangles

Above we mentioned the danger of cherry picking. This is important, whether in relation to the choice of examples or choice of theories to explain or understand the examples we choose. There is a danger, especially when students are taught by teachers without appropriate expertise in moral philosophy, that students are left with the impression that one can choose a moral theory to fit a case rather than a moral theory being an aid to evaluating or weighing different moral aspects of a situation. On this view, either one tries to see which theory fits or chooses the one which one likes best. This is an especial danger in a military context where there is always a residue (however faint) of rule, rote and orders. A parallel danger is that of presuming that there is a single determinate answer which the philosopher can give – this ignores the role of moral reasoning and judgement on the ground. Adhering to an all-things-considered approach and seeking out the morally salient features of a concrete situation for critical analysis and discussion is the antidote to this danger.

I have contended that we should not simply think of moral theories as rivals among which we must pick a winner. To explain further, utilitarianism focusses on the consequences of actions, not their independent moral quality, and asserts that we judge consequences by aggregating the balance of loss over gain and comparing the net sum of different actions or proposed courses of action. Classically, this was framed as the greatest happiness of the greatest number. In modern utilitarianism it tends to be framed as the maximising of preferences, whatever those preferences are. A utilitarian approach is reductive in the sense that it seeks to ground all moral reasoning in a single principle. However, irrespective of whether this claim can in the end be justified or not, actual moral reasoning always combines concern for consequences with concern for the rightness or wrongness of certain actions as such. To assert that certain actions are right or wrong in themselves is a feature of the deontological approach. In this approach, the right is not regarded as simply that which maximises the good, where the good is specifiable independently of the right. (For example, in utilitarianism the greatest happiness is the good and the right action is that which maximises it). In a deontological approach the right and the good are specified independently, from which it follows that there might be constraints on the pursuit of the good; an action might be 'the right thing to do' even if it doesn't maximise overall welfare, and the 'wrong thing to do', even if it does. In real moral reasoning, both features – the right and the good – are always present and in real moral reasoning there are constraints on the reduction of the goodness or badness of actions to consideration of their consequences alone. Hence the dichotomy between these forms of moral reasoning is a false one in practical moral reasoning. They are often presented as alternatives, sometimes even as alternatives that can be chosen at will or whim to fit the circumstances. This is not so. Real moral reasoning is a complex affair which necessarily comprises both.

It follows that rather than choosing between theories, my suggestion is that in learning to reason morally one should pay attention, as appropriate in any given situation, to the different aspects of moral reasoning which moral theories tend to emphasise to the exclusion of others. I have presented this in the form of a triangle, with the major theories occupying each corner. Reading across from each corner, the next triangle encapsulates the defining feature of that theory and the next across encapsulates the opposite focus to that theory. In moral reasoning one should pay attention to all of the morally relevant features of the situation by seeking an all-things-considered judgement. Practical ethics is always triangulation – it consists in applying principles and reasoning to concrete cases in conditions in which the argument is to a decision not to a conclusion, and under time and other constrained circumstances. And presentation by way of a triangle is apt too, for another reason. As we have seen, there is a tendency at the meta-ethical level to argue the merits of different theories such as utilitarianism,

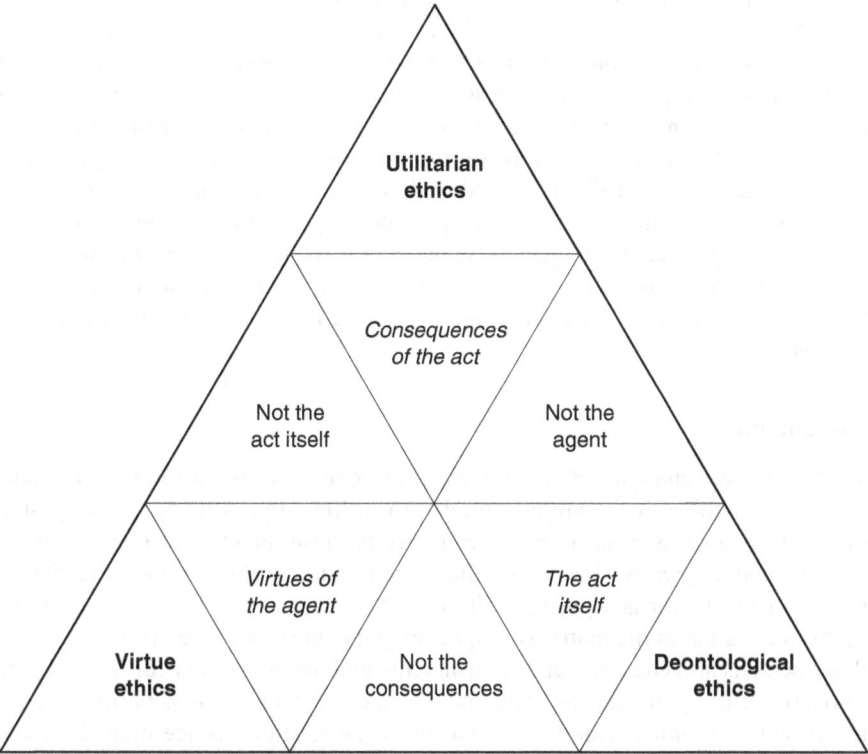

Figure 1.1 Reasoning with ethics.

deontological or Kantian ethics and virtue ethics, and the assumption is that one has to be one or the other. At the applied level the temptation can manifest itself differently as a form of cherry picking in which one is a utilitarian here, a deontologist there and a virtue ethicist elsewhere. This is clearly unsatisfactory – but the response is understandable because, outside the realm of moral philosophy as a merely academic enterprise, one does not wish to be forced to choose to be particular type of moral theorist; one wants understand how to act in concrete circumstances. The point of the triangle is to show, schematically, that if we take the three dominant ethical views and ask what is their leading feature, we can characterise all moral reasoning and decisions in the light of the weight they attach to different moral features of a situation. Although hard line adherents of each moral theory might assert that the relationship between them is disjunctive in that the defining principle of one precludes the defining feature of the others, or that (in a slightly weaker formulation) the others can be admitted only as special cases of the favoured theory, my assertion is that all moral thinking has to take account of features which are integral to each of the moral theories. We are interested in consequences and in the virtues of the agent and in following the rules. We don't have to paint ourselves into a particular theoretical corner.

One thinker who developed an approach to ethics along the lines sketched here is R.G. Collingwood (1992) who related moral theories to each other as a scale of forms comprising utility, right and duty. These overlap and only duty is truly concrete and complete in the sense that it is the only all-things-considered approach, ideally paying attention to consequences, intentions, the virtues and character of the moral agent and the historical specificity of the situation. The onus is on the historically self-aware responsibility of the choosing agent. The virtuous agent, historically informed and sensitive to circumstance chooses a course of action guided by rules, principles, consequences and so on, but through assuming the burden of judgement is hence not reducible to any of them. The key insight here is that all moral action, in the end, rests on an act of judgement by the moral agent in which that agent takes full responsibility for his or her actions.

Conclusion

Warfare is ever changing and the moral requirements of the military, both individually and collectively, are changing with it. MacIntyre (2015) has suggested that there is a crisis in military ethics partly because the character of war itself has changed so greatly. This is not the place to address this – rather, the changing character of war is a presupposition of this chapter. The remaining chapters in this book address the matter of establishing the most effective means by which those persons and educational establishments with responsibility for 'making the (modern) military moral' can train and educate the trainee troops in their charge to engage in moral reasoning. I have made suggestions concerning the most effective means of working towards this end: the approaches adopted by the other authors of chapters in this volume are in general consonant with this approach. All such authors are highly qualified academically, and over half of their number are (or have been) directly involved for many years in developing and administering ethics education in the leading military training establishments of Western democracies and thus have a wealth of knowledge and experience in the field: the remaining authors are also extremely well versed in the theory and practice of military ethics. My co-editors and I consider that, taken together, the chapters in this volume provide a clear and well-grounded guide and structure to enable their colleagues and successors to equip future members of the armed forces with the reasoning skills to deal properly with the morally problematic situations that they will encounter in future warfare. Our argument is that an approach of this sort is vital to any form of well-developed professional military ethics education.

Notes

1 I would like to acknowledge the Leverhulme Trust for funding the project 'International Network for the study of ethics education in the military' (MEEN) (F/00 181/O). This award facilitated the travel and research which made possible both this chapter and book, and the other books in the Military and Defence Ethics series.
2 I do not intend to deny, by this statement, that ethics education necessarily presupposes a moral character which is capable of such education.
3 This passage owes something to Henry Shue (2009: 320).

Bibliography

Brecher, B. (2007) *Torture and the Ticking Bomb*, Oxford: Blackwell.
Coleman, S. (2013) *Military Ethics: An Introduction with Case Studies*, Oxford: Oxford University Press.
Collingwood, R.G. (1992) *The New Leviathan*, revised edition D. Boucher (ed.), Oxford: Clarendon Press.
Collingwood, R.G. (1998) *An Essay on Metaphysics*, revised edition, R. Martin (ed.), Oxford: Clarendon Press.
Connelly, J. (2009) ' "Making Exceptions": A Response to Shue', *Journal of Applied Philosophy*, 26(3), 323–328.
Lucas, G. (ed.) (2015) *Routledge Handbook of Military Ethics*, London: Routledge.
Lucas, G. (2017) *Ethics and Cyber Warfare: The Quest for Responsible Security in the Age of Digital Warfare*, Oxford: Oxford University Press.
Lucas, G. and Demy, T.J. (2014) *Military Ethics and Emerging Technologies*, London: Routledge.
MacIntyre, A. (2015) 'Military Ethics: A Discipline in Crisis', in G. Lucas, (ed.) *Routledge Handbook of Military Ethics*, London: Routledge.
O'Neill, O. (1988) 'Ethical reasoning and ideological pluralism', *Ethics* 98(4), 705–722.
Shue, H. (2009) 'Making Exceptions', *Journal of Applied Philosophy*, 26(3), 307–322.
Singer, P.W. and Friedman, A. (2014) *Cybersecurity and Cyberwar: What Everyone Needs to Know*, Oxford: Oxford University Press.
Williams, B. (1973) 'A Critique of Utilitarianism', in J.J.C. Smart and B. Williams (eds.) *Utilitarianism for and Against*, Cambridge: Cambridge University Press.

2　Why morality matters to the military

David Fisher

Nearly fifteen years ago I was Director of the UK Defence Training Review, a major strategic review of training and education for the British Armed forces.[1] It was designed to ensure that the military were equipped with the skills needed to meet the challenges of the twenty-first century. The Defence Training Review ranged widely, looking at: how our forces needed to be trained for peace support and humanitarian operations; to operate jointly with other services and with other nations in multinational formations; and at the whole range of skills needed to cope with the changing nature of war in the twenty-first century.

The Defence Training Review introduced many reforms. This included establishing the Defence Academy which in 2012 celebrated its tenth anniversary. But what the Review did not do was to consider what training was needed by our service people in moral skills. This was the missing chapter of the Defence Training Review.

So why was moral training omitted from the review? There were no doubt a number of factors in play but I suspect there were two main reasons. The first is because at the start of the millennium there had not been a recent scandal involving unethical behaviour by the British military. Only three years later it looked very different. For there is nothing like the public outcry that attended the Abu Ghraib or Baha Mousa scandals that took place in Iraq in the autumn of 2003 to remind political and military leaders of the importance of moral training for the military.

But there was also I think a more fundamental reason at work. This is the widespread belief in contemporary society that morality is a matter of personal preference and private choices, not an area for involvement by the government; nor an area where, as good liberals, we should seek to impose our views on others. This omission of moral training was, however, in retrospect a major and serious mistake.

Importance of morality to the military

Why is morality important to the military? Above all, because morality – contrary to the widespread public perception otherwise – is not a matter of personal preferences and subjective choices. We can and do give reasons for our moral beliefs, reasons of more than local or individual appeal. Moral scepticism, in seeking to deny any such rational basis to morality, ends up unable to resist a slide towards a universal relativism in which any moral view goes and which is unable to distinguish the wisdom of Socrates from the wickedness of a Hitler or Stalin. Such ethical scepticism may sound appealing in the safety of the class-room but looks rather less appealing when we confront the consequences of human wickedness in the gulag or gas chamber.[2]

Moral rules, forged on the basis of much painful experience, provide the necessary guidelines that enable us to live well together and so flourish as human beings. So, the military, just like everyone else, need the guidance of moral rules in the daily conduct of their lives. But there are three particular reasons why morality is so important to the military.

First, because the military fight wars on behalf of the government and for the people who elect a government the morality of a war can be a matter of intense concern. Indeed, in a democracy a government that goes to war in defiance of morality may lose the consent and support of the people. Democratic armed forces may require assurance that the actions they are being asked to undertake by governments comply with morality and international law. This was evidenced in March 2003 by the request of the British Chief of Defence Staff prior to the engagement of British troops in Iraq for the Attorney General's formal assurance that their actions would comply with international law. The request set an important precedent for the future. An assurance was given in response but was far from universally accepted. The consequent persistent doubts over the legality and morality of the military action in Iraq played a significant part in under-mining popular support for the war.

The second reason why morality matters to the military is that the way they conduct wars can have a critical impact on the success or otherwise of a cam-paign. In the autumn of 2003 in the aftermath of the coalition operations to over-throw Saddam's regime there were a number of cases of abuse by US and UK soldiers of Iraqi civilian detainees. The worst incident involving British troops was the unlawful killing of Baha Mousa. Baha Mousa was a hotel receptionist arrested in the early hours of Sunday 13 September 2003. He was held in a tem-porary detention facility in Basra where he was interrogated along with eight other detainees. They were hooded, put in stress positions and deprived of sleep. They were also beaten by British soldiers. Baha Mousa was so savagely beaten that he died from the resulting injuries at 9.40 p.m. in the evening of Monday 14 September, just 36 hours after his arrest. A public inquiry into these events, chaired by Sir William Gage, was reported in September 2011.[3] Such brutal ill treatment was immoral and illegal, as the report underlined. But it also did much to undermine the strategic objectives of the campaign in Iraq. For how could we be claiming to help the Iraqi people when our soldiers were subjecting innocent Iraqis to such appalling abuse?

Moreover, it is not just the need to avoid ill-treating civilians that is crucial. As is underlined in current counterinsurgency doctrine, the protection of civil-ians may be the key to winning their hearts and minds and so contributing to the overall success of a counterinsurgency campaign. In counterinsurgency campaigns to fight well is to fight justly, to fight in accordance with the constraints of morality.

But there is also a further reason why the military need a moral code. We ask them to perform tasks that in civilian life they would be forbidden to undertake: in particular, the killing of other human beings. It is crucial for both the internal self-respect of the military and our external pride as a nation in their achievements that they are able to distinguish themselves from assassins or murderers. As a Colonel in the US Special Forces remarked, 'Our guys have got to be confident in their ability to use lethal force. But they've got to be principled enough to know when not to use it. We're not training pirates.' (quoted in Maas 2002: 55). Servicemen will return at some stage, possibly after only a few years, into civilian life where they will need to be able to look back on what they did in military service with pride rather than the disgust and shame that unlicensed killing would provoke. The military need a code of moral conduct that enables them to distinguish themselves from pirates, and see themselves rather as warriors in a noble tradition of service on behalf of the state.

That moral code may also be critical to ensuring right conduct in the heat of battle. This is well illustrated by the story of a young Marine conscript in the Vietnam War who was enraged by the death of his comrades. An officer found the youth 'with his rifle at the head of a Vietnamese woman' about to kill a noncombatant in cold blood (Osiel 1999: 23). The officer had only seconds to defuse the situation. He could have lectured the Marine on the finer principles of morality. That might have worked. But in the heat and fury of battle it probably would not have done. Instead, the officer tried a different approach. He simply said in a calm voice, 'Marines don't do that'. He appealed to the warrior code. Marines do not kill innocent women. Marines are not murderers, not pirates or assassins but warriors in a noble tradition of service to the state. The woman's life was saved. So, morality matters to the military. But what morality?

This is where just war teaching offers guidance. The just war tradition goes back to at least St Augustine in the fifth century AD wrestling with the perennial question of whether a Christian can engage in war without sin. Forged over the centuries and based on painful human experience, the tradition provides a way of structuring our thinking about war that helps guide our moral thinking.

Just war principles

Just war teaching sets out a number of tests that have to be met if a war is to be just.

Before a decision is taken to go to war we need to be satisfied that the action is undertaken with:

- just cause
- right intention
- competent authority

- as a last resort
- and the harm likely to be caused should be judged not to outweigh the good to be achieved, taking into account the probability of success.

In the conduct of a war two further tests have to be met:

- the harm judged likely to result from a particular military action should not be disproportionate to the good to be achieved by that action; and
- non-combatants should not be deliberately attacked.

Finally, the war should end in the establishment of a just peace.

All of this may appear deliberately over-prescriptive – erecting so many hurdles that war would become impossible. But the just war tradition is not a pacifist tradition. On the contrary, it recognises that war can be just and may sometimes be necessary. Nor is the tradition just a local Western (Christian) tradition. The principles are reflected in other cultures and are intended to appeal to people of reason anywhere. They seek to minimise the suffering caused by war, a matter of universal concern, not some local Western quirk. What the tradition insists on are two fundamental requirements, as simple as they are rationally compelling: Is there a just cause? And will the harm that is likely to be caused by military action outweigh the good to be achieved by that cause? In other words, is the war likely to bring about more good than harm? The just war principles are universal moral principles and are justified, like other moral principles, by their contribution to welfare and the prevention of suffering.

Applying the just war criteria does not provide some magical answer to whether or not a war is just. But they do provide a way of structuring our thinking and of ensuring that the right questions are asked and answers sought. They underpin all moral judgements about war. They also furnish the basis for our legal findings and for the international laws of war.

But what I want to consider in this chapter is what else is needed to ensure the just conduct of war.

Moral rules not enough

The just war principles provide us with a moral rule book for war, a road map for the campaign. A rule book is certainly necessary for the ethical behaviour of those involved with war. But our aim is not simply to ensure that our political leaders and service people are well versed in just war teaching. What matters is that they behave justly. The challenge is to ensure that the right decisions are taken before, during and after war, even amidst all the passions and fury this may arouse. Moral principles alone will not suffice to achieve this. A moral rule book is necessary to guide our actions but it is only part of what we need. So, what else is needed?

Moral principles provide, as I have said, the necessary guidelines for how we can live together well in communities and so flourish as human beings. But we also need to have acquired appropriate habits and skills to enable us to put those principles into practice in our daily conduct. This is the role played by the virtues. Virtues are those habits of thought, feeling and action that enable us to choose what is right and then to act rightly, despite the many pressures there may be to do otherwise. The virtues are crucial to moral conduct generally but they are particularly important in the area of war where the pressures to act unjustly are likely to be particularly acute.[4]

Virtues

So, what virtues are needed to ensure the just conduct of war? In traditional medieval thought, based on Aristotle, four virtues are identified as pre-eminent or 'cardinal' in our moral lives. They are: practical wisdom, courage, self-control and justice. Let us consider each in turn.

The just war principles are inevitably of a general nature. We need, therefore, help and guidance in relating the general to the particular: to recognise which principle applies to which particular situation, to discern how it applies and what action it requires. Such guidance is furnished by the virtue of practical wisdom. That may sound academic and remote. But practical wisdom is very far from being either academic or remote. Practical wisdom is 'a habit of sound judgment about practical situations' (Geach 1970: 160). Such a habit of sound judgement about practical situations is what distinguishes a good military commander from an indifferent one, or a jobbing politician from a statesman. Such a habit of sound judgement is what is needed on the battlefield when time is short and instant decisions are required, with many lives at risk. Practical wisdom is essential to ensure the correct application of the just war principles to the messy, complex and shifting realities of the battlefield. It is the virtue that we need above all to ensure the just conduct of war.

Military practical wisdom was for Aquinas a virtue to be exercised chiefly by the general commanding an army. Generals directing warfare at the strategic and theatre levels of command need, by careful training and experience, to have acquired and exercised practical wisdom. They will need this to enable them to provide sound advice to their political leaders and to make right decisions on the battlefield about how the war and its individual battles are to be conducted. They will need this to enable them to assess correctly whether the course of military action proposed will bring about more good than harm.

It is a feature of modern warfare that critical decisions on the conduct of war are being devolved to ever lower levels. Practical wisdom needs to be exercised at the tactical level, often by service people of junior rank. The 'strategic corporal' can be responsible for taking life and death decisions. It is he or she who may have to decide whether the passenger in the car at the checkpoint on the Afghan roadside is a dangerous insurgent or an innocent citizen to be protected by the principle of non-combatant immunity. The modern battlefield demands of all its participants an increasing range of skills, with activities capable of shifting within the space of three city blocks from war fighting through peace support operations to humanitarian relief – the so-called 'Three Block War'. Practical wisdom needs to include an ability to adapt rapidly and respond appropriately to fast-changing situations. Since the decisions may have to taken in split seconds, the exercise of practical wisdom needs to have become, through long training and practice, second nature, a deeply ingrained habit of sound judgement.

Practical wisdom is exercised at a variety of levels, ranging from that of the statesman deciding whether or not to go to war to that of the ordinary soldier deciding how to react at a vehicle checkpoint. The application of the just war principles to the messy, complex realities of daily life in peace and war needs, above all, the virtue of practical wisdom. But the just conduct of war also requires the other cardinal virtues of courage, self-control and justice.

Courage is the state of character that we need to enable us to persevere in the face of difficulty and danger, not allowing fear to obscure our judgement of what needs to be done. Physical courage is the much-prized virtue of the warrior on the battlefield. But service people may also require moral courage to stick to their principles despite pressures from their comrades or superiors to do otherwise.

It was moral courage that was lacking in the officers – including the doctor and the padre – who failed to do anything to stop the savage beatings and other abuse by British soldiers of Baha Mousa and the other civilian detainees held in the Temporary Detention Facility (TDF) in Basra in September 2003. Sir William Gage, who chaired the public inquiry, expressed 'concern about the lack of discipline and lack of moral courage to report abuse within 1 QLR (Queen's Lancashire Regiment)'.[5] He criticised the commanding officer, Colonel Mendoca who 'ought to have known what was happening in the TDF'.[6] He criticised the padre who

> must have seen the shocking condition of the detainees, and the deteriorat-ing state of the TDF. He ought to have intervened immediately, or reported it up the chain of command but, in fact, it seems he did not have the courage to do either.[7]

The army doctor who claimed at the enquiry not to have noticed the 93 external injuries that led to Baha Mousa's death appeared in June 2012 before a tribunal of the General Medical Council and was charged with dishonesty and miscon-duct.[8] It is a key recommendation of the Gage report that we need to devise better ways to teach moral values in general and moral courage in particular to the military.[9]

Moral courage is the virtue that a service person needs to challenge unac-ceptable behaviour and to disobey orders to carry out immoral or illegal actions. It is a difficult virtue to practise, particularly in an institution, such as the armed services, that rightly also prizes the virtues of obedience and loyalty. The exer-cise of moral courage at times appears to conflict with that of these other virtues.

The difficult challenge required to exercise moral courage is illustrated by the agonising of the US soldier, Adam Winfield, before he agreed to blow the whistle against his Staff Sergeant, Calvin Gibbs. From his appointment in November 2009, Gibbs had led a 'Kill Team', of which Winfield was a member, that had murdered innocent civilians in Afghanistan. In a letter to his father after the death of one such innocent, Winfield wrote:

> Pretty much the whole platoon knows about it. It's OK with all of them pretty much. Except me. I want to do something about it. The only problem is I don't feel safe telling anyone. The guy who did it is the golden boy in the company who can never do anything wrong and it's my word against theirs.[10]

Misplaced loyalty to his comrades and fear of the consequences of speaking out inhibited Winfield from doing what he knew was the right thing. Such misplaced loyalty had similarly led to the 'wall of silence', criticised by the judge in the Baha Mousa court martial that had preceded the public inquiry, from the officers and soldiers, all of whom had declined to give evidence against their comrades involved in the abuse of Iraqi detainees.[11]

For service people, operating in a culture where importance is given to obeying orders and loyalty to their comrades, it presents a particularly difficult challenge to recognise the occasion when an order should be disobeyed, and to act accordingly, despite the many countervailing pressures. This requires not just the exercise of moral courage but also of the virtue of practical wisdom – 'a habit of sound judgment about practical situations'. If abuses such as those that occurred in the Baha Mousa case are to be prevented in the future much higher priority needs to be given within the military to the virtue of moral courage. This higher priority will need to be demonstrated not just by what is taught in training schools but much more by the behaviour that is encouraged or discouraged each day in the barracks and on the battlefield. This will necessitate not only the better training recommended by the Gage Report but will also require a major culture change within the armed services.

As well as courage, service people need the cardinal virtue of self-control. They need to exercise self-control if they are to achieve and maintain the physical fitness required to undertake their duties, neither eating too much nor too little to sustain their physical strength. Self-control is, like practical wisdom, a virtue that is needed in support of the exercise of other virtues. It is the character trait that helps a man never lose his head, whether through anger or fear or lust for pleasure and so always to exercise a dispassionate judgement of what is the right thing to do. Self-control is a key virtue on the battlefield.

It is one of the many virtues lacking in the soldiers in the Basra TDF who allowed their anger at an unfounded rumour that the detainees had been connected with the earlier murder of three Royal Military policemen to inflame their actions. Self-control is a key virtue on the battlefield. It is also a virtue that those outside the military profession require. This includes our political leaders and their advisers who need to take decisions on war and peace without allowing anger, greed or other passions to obscure their judgement.

Finally, and crucially, our politicians and service people need to learn and practise the virtue of justice, showing concern for others and learning to respect and promote their good. The just war principles set out the criteria to be met if war is to be just in its inception, conduct and conclusion. To enable us to meet this challenging demand requires that we learn and practise the virtue of justice, respecting and promoting the welfare of others, even those who may be our enemies.

So, service people, like ordinary citizens, need the four cardinal virtues of practical wisdom, courage, self-control and justice. What other virtues do they need? The British Army's statement of its Core Values and Standards curiously does not set out values or standards. But what it does do is delineate the qualities, the virtues – a word that had perhaps become too unfashionable to use in the title of a contemporary military document – that are required of our soldiers. These are: selfless commitment, courage, discipline, integrity, loyalty and respect for others.[12]

I have already considered courage, as well as the respect for others which underlies the virtue of justice. Selfless commitment is about the individual placing the interests of the group within which he serves above his own interests. Discipline is necessary to ensure the prompt execution of orders on which the lives of others may depend, particularly in the heat of battle. Such obedience is critical in the military context, although it should not be blind, unthinking obedience. The obedience needs to be reflective, able to discern and, if necessary, decline orders that are illegal or immoral. Integrity is about truthfulness and honesty. It can be critical to the saving of lives which can be put at risk by inaccurate information. It requires consistency between principle and practice: meaning what one says, saying what one means and doing what one says. Integrity is important for a leader to help him earn the trust of those he commands.

Finally, loyalty to one's comrades and to the military unit in which one serves is a virtue much needed on the battlefield. It is the virtue that drives the service person to go back and rescue a comrade, even though this may put his own life at risk. Loyalty needs, however, as I have stressed, to be properly directed and not so narrowly focussed as to be misplaced.

The British Army's Core Values provide a good summary of most of the qualities needed by the military. But much greater emphasis needs to be given to moral courage. There are also two key additions that need to be made: the cardinal virtues of self-control and practical wisdom which are as crucial to military, as non-military, life. Indeed, they are the virtues that underpin all the others and are central to our moral life.

Teaching justice

In his introduction to the Army's own internal report into the Baha Mousa case – the Aitken Report – the then Head of the British Army, General Sir Richard Dannatt, noted that, 'we need to find better ways to inculcate our core values'.[13]

Since the events in Basra improvements in ethical training have been made, but the failures in Basra, as Dannatt underlined, point to the need for a radical over-haul of the way values and standards are taught within the military. The need to find better ways to teach morality to the military is also a key recommendation (accepted by the UK Government) of the Gage report following the public inquiry into those incidents.

Morality needs to be given much higher priority and salience within the military curriculum. Ethical teaching should include not just the laws of war, such as the Geneva conventions, of which our soldiers seemed so sadly unaware. It should also embrace the wider body of just war teaching, whose practical wisdom and insights should be expounded not just to officers but to all ranks, reflecting the way responsibility is devolved to ever lower levels in con-temporary conflicts. Just war principles need to become staple fare in the class-rooms of both officers and other ranks.

But morality cannot only be taught in the classroom. Morality needs to be expounded not as an abstract set of principles but as a practical guide to daily living, taught by precept, personal example and practice. Examples of virtuous behaviour need to be studied from both living and historical exemplars. This should also include, as the Gage Report recommended, examples of vicious behaviour to avoid any risk of complacency.

The personal example set by the military leader is, of course, crucial. It is also important to recall that morality is taught not just by what we say and do but what we fail to say and do. It was noticeable how Lynndie England, one of the participants in the Abu Ghraib abuses, argued in her defence that she had assumed the abusive practices were acceptable because no-one had told her otherwise, 'When we first got there, we were like, what's going on? Then you see staff sergeants walking around not saying anything (about the abuses). You think, OK, obviously it's normal'.[14]

The just war principles need to be expounded not just in our classrooms but embedded in the daily practice and experience of the barracks and battlefield. Moral training needs to become an integral part of the very fabric of service life. For it is not enough that our service people have a theoretical understanding of the conduct of just war. What matters is that they behave morally. Learning the rules and codes of war is necessary for the ethical behaviour of those involved with war. But rules alone are not sufficient to ensure that the right decisions are taken before, during and after war. Just as an artillery officer needs both an understanding of the science of gunnery and to be daily drilled in the tasks of artillery warfare, so our service people need to be equipped with both the theory and practice of morality. They need moral *education* so that they know and understand the principles of morality in general and those relating to the conduct of war in particular. But they also need to be *trained* so that they have the appro-priate character traits and skills to apply the principles in practice to the diverse challenges and tasks with which they may be faced, even when under pressure and with time running out. It is important too that they are able rapidly to adapt their behaviour to the shifting battlefield, as tasks change from those of conflict to those of peaceful reconstruction, a shift in behaviour that the soldiers who beat Baha Mousa to death failed to achieve.

All those involved in decisions about peace and war, from the highest level down to the ordinary service person, need to have been educated and trained to confront the moral challenges with the appropriate beliefs, desires and feelings and to have acquired the necessary skills and practical wisdom to enable them to behave morally Our service people need, in other words, to have been schooled and practised in the virtues so that ethical conduct becomes for them second nature, as deeply engrained as habits of thought and action as are the drills with which they wield their weapons.

I started this chapter recalling the UK Defence Training Review. I was struck during that review by what a young corporal who had taken part in the successful British intervention in Sierra Leone in September 2000 told the review team. He told us: 'This is the only fire fight I've ever been in. This company is a very young company and none of us had ever experienced it before. But when the battle started *the training just took over*'.[15] He was, of course, referring to his training in weapons drills and firefighting. But what he said equally applies to moral training. If we are to ensure that abuses, such as those that happened in the British TDF in Basra or under US command in the prison at Abu Ghraib, are not to recur, we need to take ethical training as seriously as weapons training. Like the corporal in Sierra Leone, we need the moral training to '*just take over*'.

Notes

1 *Modernising Defence Training – Report of the Defence Training Review* (Ministry of Defence, 2001).
2 The objections to moral scepticism are explored in more detail in David Fisher, *Morality and War – Can War be Just in the Twenty-first Century?* (Oxford and New York: Oxford University Press, 2011, pbk., 2012, especially Chapter 2 'Whose Justice? Which Rationality?'.
3 *The Report of the Baha Mousa Inquiry*, The Rt. Hon Sir William Gage (Chairman) (London: The Stationary Office, 8 September 2011).
4 The role and importance of the virtues in military life are discussed in more detail in Fisher, *Morality and War*, especially Chapter 6 'Virtues'.
5 *Report of the Baha Mousa Inquiry*, Volume III, Summary of Findings, paragraph 203.
6 Ibid. paragraph 251.
7 *Report of the Baha Mousa Inquiry*, paragraph 112.
8 Derek Keolloh, 'Baha Mousa death: army doctor accused of cover-up, online at: www.guardian.co.uk/world/2012/jun/11/baha-mousa-doctor-accused-coverup.
9 *Report of the Baha Mousa Inquiry*, Vol III, Part XVII, Recommendations, Recommendation 58.
10 Chris McGreal, '"Kill Team" US platoon commander guilty of Afghan murders', online at www.guardian.co.uk/world/2011/nov/11/kill-team-calvin-gibbs-convicted, 11 November 2011.
11 Aitken Report, *An Investigation into Cases of Deliberate Abuse and Unlawful Killing in Iraq in 2003–4* (London: UK Ministry of Defence, 25 January 2008), 24.
12 The 'Values and Standards of the British Army' can be viewed at: www.army.mod.uk/documents/general/v_s_of_the_british_army.pdf.

13 Aitken Report, 1.
14 Emma Brockes, 'What happens in war happens', an interview with Lynddie England, *Guardian*, 3 January 2009.
15 *Report of the Defence Training Review*, 5.

Bibliography

Aitken Report (2008) *An Investigation into Cases of Deliberate Abuse and Unlawful Killing in Iraq in 2003–4*, 25 January, London: UK Ministry of Defence.

Fisher, D. (2011) *Morality and War – Can War be Just in the Twenty-first Century?* Oxford and New York: Oxford University Press.

Gage, Rt. Hon. Sir William (Chairman) (2011) *The Report of the Baha Mousa Inquiry*, 8 September, London: The Stationary Office.

Geach, P. (1970) *Virtues*, Cambridge: Cambridge University Press.

Keolloh, D. (2012) 'Baha Mousa death: army doctor accused of cover-up, *Guardian*, 16 December 2012, online at: www.guardian.co.uk/world/2012/jun/11/baha-mousa-doctor-accused-coverup.

Maas, P. (2002) 'A Bulletproof Mind,' *New York Times Magazine*, 10 November 2002.

McGreal, C. (2011) '"Kill Team" US platoon commander guilty of Afghan murders', *Guardian*, 11 November, online at www.guardian.co.uk/world/2011/nov/11/kill-team-calvin-gibbs-convicted.

Ministry of Defence (n.d.) 'Values and Standards of the British Army' online at www.army.mod.uk/documents/general/v_s_of_the_british_army.pdf.

Ministry of Defence (2001) *Modernising Defence Training – Report of the Defence Training Review*.

Osiel, M. (1999) *Obeying Orders: Atrocity, Military Discipline and the Law of War*, New Brunswick, N.J.: Transaction Publishers.

3 Military ethics and the importance of cultural competency

George R. Lucas Jr

The mistaken burning of prisoners' copies of the Qur'an by American military personnel at Bagram AFB, Afghanistan in February 2012, and the tragic, violent social upheavals that followed in its wake, constitutes an unfortunate example of the crucial importance of regional knowledge and cross-cultural competence in the conduct of counterinsurgency. It is beyond exasperating to perceive how a single, symbolic cultural blunder of this magnitude can undo months and years of careful diplomacy and capacity building, and threaten the meaningful legacy of the many who have lost their lives in this conflict. The American military's cultural deficit has hardly gone unnoticed, but it has proven difficult in the extreme to assure that such knowledge and competence is attained uniformly and adequately by military personnel (as this terrible incident likewise illustrates), as well as to determine how best to provide such education in essential regional knowledge, and to assure such cross-cultural competence is readily available in theatres of conflict, especially to commanders in the field.

That American military personnel and NATO allies in ISAF and, even more, American and coalition forces in Iraq lacked such knowledge in a manner that was undermining counterinsurgency efforts was first recognized by then-Lieutenant General David H. Petraeus of the U.S. Army as early as 2005. Petraeus learned this lesson first hand, while in command of coalition troops in the Sunni-dominated Anbar Province in Iraq in the early days of that conflict. His efforts to better understand, reach out to, and cooperate with local and regional tribal leaders in establishing peaceful control over the region, however, ran afoul of prevailing American policy early in the Iraq war, that dead-set against any reconciliation or accommodation with those who enjoyed rank or privilege of any sort during the regime of Saddam Hussein. Petraeus was summarily returned to the U.S. upon completion of his tour of duty, and posted to Ft. Leavenworth, KS to serve as commander-in-chief of the U.S. Army's Training and Doctrine Command (TRADOC). For a capable and ambitious military leader during a critical time of armed conflict, this re-assignment constituted a de facto exile.

Petraeus made good use of his time at Leavenworth, however, propounding his radical views on the pre-eminent importance of enhanced regional and cultural knowledge, and weaving them into the revised doctrine of counterinsurgency warfare (COIN), subsequently issued under his editorship in December

2006 in a newly revised U.S. Army and Marine Corps field manual guiding such operations. By this time, Petraeus's views had returned to favour, and he was posted to Iraq essentially to redeem, through these methods, the floundering U.S.-coalition peace-keeping efforts there, which had all but collapsed into wholesale civil war.

The most public face of that recovery at the time was the so-called 'surge' in the number of coalition troops deployed as ground forces, and the redistribution of those forces from safe havens in military compounds and the so-called 'Green Zone' in Baghdad, to highly visible local village patrols in virtually every region of the country. But Petraeus also put a key second component of his counterinsurgency strategy into place at this time in Iraq as well, with somewhat less fanfare, deploying five new 'Human Terrain Teams' to work alongside Brigade Combat Teams (BCTs) in key areas of the country in early 2007. Each team consisted of specially chosen military and civilian personnel, recruited for their specific regional and cultural expertise and experience, and trained (during a four-month intensive programme at Ft. Leavenworth) to collaborate effectively with multiple BCTs at once, in order to enhance the effectiveness of military and security ground operations, largely by ameliorating and reconciling (rather than inadvertently causing) cultural conflicts of the kind that the mishandling of the sacred texts in Afghanistan so vividly illustrates. This programme, known as the 'Human Terrain System'(HTS), was thus the U.S. Army's ambitious attempt to increase dramatically the degree of regional knowledge, cultural awareness, and anthropological expertise required for successful military counterinsurgency operations, and to recognize, anticipate, and head off in advance the kinds of cultural blunders and misunderstandings (such as the ritual mishandling of copies of the Qur'an) that might otherwise lead to disaster.

While quietly implemented during its first two years, the programme gained considerable notoriety with those initial deployments of HTS teams to Iraq and Afghanistan in 2007. A good deal of the controversy came from perceived efforts by the U.S. military to recruit academically trained anthropologists and other social scientists to anchor each team's cultural knowledge and expertise. Anthropologists, in particular, took umbrage at what they characterized as an attempt to leverage disciplinary knowledge and professional expertise for what they viewed as illicit purposes, especially in the conduct of an unpopular and purportedly illegal armed conflict in Iraq. Thus, discussions of the professional probity of anthropologists, in particular, cooperating in the conduct of hostilities quickly became hopelessly mired in the broader political disagreement over the legitimacy of the American war efforts themselves.

The work was also inherently dangerous, entailing risks not unlike those faced by journalists, health care workers, and humanitarian relief and development personnel. Charges of improper and incompetent administration of the programme levelled by its critics increased substantially in the wake of the death of three anthropologists, one former Army enlisted personnel working in Iraq, and two developmental anthropologists working with HTS teams in Afghanistan in 2008. The resulting, widely publicized furore over what came to be called 'military anthropology', was the topic of my book, *Anthropologists in Arms*, published in 2009.

The wider controversy, and the book devoted to its analysis and examination, treated many topics beyond HTS itself: questions of professional ethics and the nature of profession malfeasance, for example, as well as the tensions inherent in the role of scientist and citizen, especially during time of war. Membership in, and governance of, voluntary professional communities themselves, particularly as regards professional jurisdiction and responsibility for individual malfeasance, was another important topic raised in this controversy, as were the conceptual differences (and often morally disturbing similarities) between routine field work in the social sciences generally, and outright espionage. Informed consent, and protection of human research subjects also loomed large, especially in the context of an environment where some of those 'subjects' might be found to constitute legitimate enemies or adversaries, for whom it would become the task of military, intelligence, and security forces themselves to capture or kill.

Ironically, my analysis also revealed that the actual administration of the HTS programme (including recruiting, training, deploying, and compensating the academic social scientists involved) had been outsourced almost from the beginning as a private contract to British Aerospace (BAE Systems, Inc). Thus, anthropologists and others participating in this programme were, in fact, classified as private military contractors (in the employ of BAE Systems, rather than the U.S. military). This quickly embroiled HTS itself in the ostensibly distinct public and political debate over the propriety and efficacy of private military contracting, a novel dimension of what I termed at the time 'post-modern warfare'. Indeed, the crescendo of cries of outrage over HTS was unfortunately timed to coincide with public anguish following the debacle of armed Blackwater security contractors seeming to fire indiscriminately, and killing and injuring some 17 Iraqi civilians at Nisoor Square, Baghdad in September 2007).[1]

An important subtext of the book, woven throughout this wider public debate, however, consisted of a question that was never adequately brought to the fore or thoroughly examined: namely, how should military forces, and programmes of professional military education and leadership development, go about the task of preparing their personnel to operate effectively (including both 'justly' and 'respectfully') in dramatically unfamiliar cultural environments? If we accept, on all the various grounds adduced above, that it is properly the responsibility of military, intelligence, and security forces to exhibit adequate knowledge of the 'human terrain' within which they are obliged to operate, and to manifest consistently 'good cultural practices' while operating in those environments, how are those vital capacities best engendered, imbued, inculcated, and maintained? It is, transparently and pre-eminently, *this* question that the unfortunate Qur'an incident at Bagram AFB so urgently underscores. And this is not a question that is readily or easily answered.

It is, however, a question properly considered within the framework of the present volume, inasmuch as provision of this cross-cultural competence might properly be regarded as a 'moral responsibility' of professional military education generally, while the content of that expertise also has an ethical component. Due to absent sufficient cultural awareness and regional competence, military professionals are both unable to do their job of providing peace and security properly, and are at grave risk of exacerbating (rather than ameliorating) conflict and of harming (rather than safe-guarding) those whom they are sent to protect. Thus, against the valuable background of previous contributions made in their previous publications by members of the Military Ethics Education Network (MEEN) to our understanding of effective moral education and character development of military personnel, it behoves us to consider whether there are better or worse methods of going about the important task of developing regional knowledge and effective cultural expertise for military forces.

In that sense, while it attempted a rapid response to an urgent need, HTS also represented one of the military's persistent failings in seeking a 'quick fix' to a complex problem. HTS was designed as an 'off-the-shelf' program: lacking relevant expertise, the organization simply tried to go out and purchase the relevant experts in the open market, rather than having invested the time and care to cultivate that expertise within its own ranks. There are, of course, drawbacks to this approach of a cultural sort as well: the newcomers are seen as strange aliens, not a part of the 'warrior culture' in this case, and threatened with misunderstanding and marginalization. Those feelings can be mutual, and lead to the kind of disaffection that apparently led some HTS members to quit prematurely and criticize the programme. The newcomers also present a practical obstacle to effective battlefield integration, much as NGOs trying to operate amidst armed conflict sometimes present a risk to themselves and others. This is hardly desirable in a battlespace already overpopulated with increasing varieties of non-military personnel, often acting without licence or coordination. Thus, the 'quick-fix' of HTS quickly came to represent a different kind of problem: how do regular military personnel work effectively with academics who know nothing of combat?

A second approach to providing 'human terrain' resources to military personnel was actually tried first, with some success. 'Crash courses' in cultural education were added to pre-deployment training in so-called 'reach back' centres at Marine Corps University (Quantico) and Army Staff and Command College (Ft. Leavenworth). Prospects for deployment were regaled with smart cards and interactive CD-ROMs, distilling presumably essential cultural knowledge into a handy pamphlet or fact sheet that could be carried on one's person as a quick reference guide. 'Reach-back' centres also served as headquarters for regional and cultural experts who taught these courses and compiled and distilled the essential knowledge, while providing an accessible database of cultural information that could be accessed remotely by commanders in the field via secure internet and satellite connection. Live consultation of brigade teams with cultural experts back home via satellite and VTC was also an option.

This initial approach to the cultural knowledge deficit predictably appealed more strongly to academic experts than an intensive programme of rigorous basic training followed by actual deployment in combat would have done. And, to be fair, such a solution also addressed the uncomfortable logistical problem described above, by avoiding the overpopulation of the battlespace with a variety of largely untrained and unprepared non-combat personnel. It is not a bad solution, but it is likewise not a very good one. Assembling all this expertise for remote consultation was probably too little and too late to adequately address the kinds of problems that General Petraeus had witnessed during his initial tour in Iraq. Academic and professional authorities would argue that complex, nuanced cultural knowledge is not easily, or even appropriately, distilled or 'boiled down' on a wallet card for use by a relatively untrained new recruit. And brigade commanders themselves often expressed a desire for immediate, on-site advice – an 'Angel on the shoulder', as they put it – in lieu of a 'guide on the side', remotely accessible only by advance appointment. The kinds of practical problems being encountered required a more rapid response time than this leisurely (and safer) academic model could provide.

The obvious solution for the long term, of course, is to routinely strengthen the preparation of officers and enlisted personnel in cultural and regional knowledge, alongside foreign language skills. This is very likely the optimal solution in terms of cultivating the required expertise within the ranks of the organization itself, but it is also the most time-consuming and expensive. And it is not sensitive to rapid change: by the time we had educated military officers with competence in Urdu or Pashtun, the focus of both conflict and urgent need would have moved on – now we are more concerned with what is happening in Syria or Libya rather than in Afghanistan or Iraq, although the latter two countries are still, to say the least, a very long way from being at peace.

Critics of this 'long-term investment' approach to cultural expertise would likely point to the legions of Russian language and culture experts in military and intelligence, as well as in academic circles, left high, dry, and bereft of useful employment following the end of the Cold War. (And, lest the irony be lost, no sooner do we abandon them and berate the 'outmoded' model, than their area of expertise once again flares to the surface as political tensions with the Russian Federation arise.) Our response time, on this model, is perpetually lagging behind the present needs, and utterly out of phase with the pace and direction of cultural transformation.

The U.S. Air Force, cognizant of these myriad problems, has developed a different approach to regional and cultural education through a programme dubbed 'C-cubed' (C^3) – 'Cross Cultural Competence'. For the reasons adduced above (proponents of this approach argue), it is not efficient or feasible to provide rigorous and thorough cultural knowledge of a specific sort, certainly not throughout the organization, and to each and every individual member of it. Instead, the Air Force philosophy is to educate military personnel to adapt effectively, and interact intelligently across a range of unfamiliar cultural situations, to develop a kind of 'multi-cultural sensitivity' – and leave the specific languages and knowledge of regional customs to locals and to a few experts within the organization.

Each approach has its own inherent limitations. The HTS programme represented an ambitious and radical approach to addressing the cultural deficit.

Long-term assessment of the program's effectiveness has yet to be accomplished. The other approaches are both more tried and true, and perhaps pedagogically satisfying, but, as noted, are slow, inefficient, and unresponsive to emerging needs. The Air Force C^3 model offers perhaps the most promise of effectively sensitizing and equipping military personnel, quickly and with reasonable efficiency and effectiveness, to operate in unfamiliar cultural terrain. It remains an open question whether such an approach, adopted through allied and coalition militaries, would be sufficient to minimize risks of a Bagram-like cultural incident of a sort that otherwise, and without warning, can bring a well-intentioned relief, development, and nation-building effort to the brink of despair.

Note

1 This is no longer the case. Following the Nisoor Square incident, the controversial practice of granting legal immunity of foreign civilian contractors under local domestic law was repealed in Iraq. For their protection, HTS employees were henceforth converted to regular civilian civil service positions in the U.S. Dept of Defense upon completion of the training in Ft. Leavenworth.

4 Solving the military moral bystander problem with ethics instruction

Peter Bradley and Allister MacIntyre

Introduction

Observers of warfare at the turn of the twenty-first century might expect that technological advances such as precision-guided munitions, unmanned vehicles, war robots and the like would make modern conflicts more surgical and perhaps less violent. Indeed, this might be so for drone pilots stationed many thousand miles away from their actual battlefield, but for soldiers on the ground in face-to-face combat with their opponents, war appears to be as messy and violent as ever. Soldiers in operations still face the daunting challenge of juggling their finite physical, mental and moral resources to meet the competing demands of ensuring the safety of their comrades, themselves and nearby noncombatants, while also pursuing their ultimate objective: achieving the mission. Sometimes the concerns for personal safety, fatigue, frustration and the many other psychological challenges become too much for soldiers, and their reactions extend beyond the legal and ethical limits of acceptable behaviour. In earlier times violations might have passed unnoticed, at least for a while, but with public and social media as pervasive as they are today, concealing these transgressions is virtually impossible, and horrific results can occur, as we have seen (e.g. Abu Ghraib). Most of the Western military forces involved in the recent wars in Iraq and Afghanistan have had cases of soldiers[1] committing acts which violate the laws of armed conflict and professional codes. What follows is a small sampling of cases.

In September of 2003, British soldiers searched a hotel in Basra and detained seven men they suspected of being loyalists of the former Iraqi regime. Over the course of the next 36 hours, the captors abused their detainees to the point that one of them, a hotel worker named Baha Mousa, died from his beatings. A post mortem examination conducted several days later found '93 separate surface injuries on Baha Mousa's body' (Gage Volume 1 2011: 6). Charges were brought against seven British soldiers, including the commanding officer, as a result of this incident. However, only one of the soldiers, a corporal, was convicted after pleading guilty to inhumane treatment.

A few years later, in March of 2006, four U.S. soldiers, fuelled by alcohol and boredom, broke into the home of an Iraqi family to rape the 14-year-old daughter that one of the soldiers had been watching for some time. While one soldier killed the parents and a younger daughter, others raped the 14-year old, killed her and then set her body on fire (Frederick 2010: 264–268). The four were convicted for their crimes and sentenced to lengthy prison terms.

On another battlefield, this one in Afghanistan, a Canadian captain fired two rounds into a mortally wounded Taliban fighter in October of 2008. The captain was charged with several offences and brought before a military court, but because he didn't testify at his trial, the public never learned why he shot the incapacitated enemy fighter. In the end, the captain was convicted of disgraceful conduct and released from the Canadian Forces (Friscolanti 2010: 20; Friscolanti and Geddes, 2010).

The cases mentioned above and others like them illustrate two types of professional failure in the military, both of which have drawn very little interest on the part of researchers or commentators. The first is what we call the 'moral bystander', when another soldier – or perhaps several – witnessed unit comrades violate laws or professional codes, but failed to stop the transgressions as they were happening. The second is the absence of what has been described in organizational studies as a whistle-blower, someone who was present at the offence – a bystander perhaps – or someone who heard about the transgression later, and did not report the wrongdoing to proper authorities. Because soldiers were reluctant to come forward in the U.K. case mentioned above, 'some soldiers who had abused the detainees [have] not been charged ... because there is no evidence against them as a result of a more or less obvious closing of the ranks' (Gage Volume 1 2011: 1). In the U.S. example, some platoon mates of the four perpetrators knew of the offences, but were reluctant to report what they knew. In the Canadian case, three other Canadian soldiers were with the captain that day, but they did not report his wrongdoing. It was only when an Afghan interpreter came forward that an investigation was launched and legal action taken. In all three cases and perhaps many others just like them, soldiers neglected their moral and professional military obligations and stood by while their mates committed misdeeds. In addition, these bystanders, and possibly others who were aware of the misconduct, compounded the professional failure by not reporting the violations to a recognized authority.

In this chapter, we wish to examine the feasibility of employing ethics instruction to reduce the moral bystander problem and encourage responsible whistleblowing in the military. An earlier Ashgate collection showed that Western nations have allocated considerable resources to military ethics training in recent years (Robinson *et al.* 2008) and we wish to add to that work by showing how instruction on the bystander effect and healthy whistleblowing can enhance military professional development.

Our chapter has seven parts. First, we summarize research on the bystander effect which explains why some individuals passively stand by while their colleagues engage in illegal or unprofessional behaviour. Second, we review whistleblowing research which demonstrates why individuals can be reluctant to report the transgressions of their workmates. Third, we show how turning a blind eye while unit mates violate laws and professional codes is inconsistent with the tenets of military professionalism. Fourth, we consider the cognitive and motivational challenges bystanders and whistle-blowers face, along with a number of the common errors and biases which can keep them from making sound decisions. Fifth, we discuss the pros and cons of trying to solve the bystander/ whistle-blower problem with selection tests. Sixth, we review ethics training effectiveness research for insights on how training may help. Seventh, we offer some suggestions on what soldiers should be taught to equip them to react properly to the wrongdoing of comrades.

Bystander research

When reflecting on the cases described in our introduction, some readers might take comfort in the belief that they would have somehow tried to find a way to intervene had they been present. Yet, there is a body of research suggesting that this is not the case. So far there is very little research on this type of moral indifference in the military, but an abundance of examples of brutality, rape, murders, and cruelty in general society shows that these behaviours are pervasive. And the passivity of the witnesses who have been present during these acts of depravity is also well documented.

One of the most widely known and cited cases of this nature is the attack and murder of Kitty Genovese in March 1964. Kitty was beaten, raped and stabbed while many neighbours purportedly not only witnessed the assault, they did not call the police, get involved, or attempt any sort of rescue effort. Although the details of the case have been disputed over the years (e.g. Cook 2014), the death of Kitty Genovese stimulated a field of psychological study (e.g. Darley and Latane 1968; Latane and Darley 1968, Latane and Nida 1981; Cacioppo *et al.* 1986) into what has come to be known as the *bystander effect*. The bystander effect presents a bit of a paradox because we expect that there will be strength in numbers. Yet, the research demonstrates that the likelihood of witnesses intervening to help, when someone is experiencing distress, decreases as the number of people present during the situation increases.

In their review of the literature up to that point, Latane and Nida (1981) identified three possible reasons for bystander inaction. The first reason, referred to as *audience inhibition*, stems from a motivation to avoid embarrassment. The perceived risk of negative evaluation increases as the number of observers increases. The second reason, *social influence*, is a consequence of the ambiguity that characterizes many emergency situations. When people have difficulty understanding the dynamics of a situation, they look to others present to see how they are reacting. When they notice that the others are not taking action, they will either re-evaluate the circumstances as less critical, or accept inaction as the expected norm of behaviour. The final reason has been labelled *diffusion of responsibility*. The presence of others means that the psychological costs associated with inaction become shared and this allows the responsibility to be shifted from individuals to the group. When everyone is responsible, no one feels responsible and the likelihood of intervention becomes less likely (Latane and Nida 1981). While diffusion of responsibility suggests that people will be less likely to help because the responsibility to render aid becomes a shared burden, Cacioppo *et al.* (1986) contend that *confusion of responsibility* might cause some bystanders to not intervene from fear of being seen by other observers as being involved in the misconduct and responsible (either partly or entirely) for any harm to the victim.

Is it possible that the bystander effect could be used as a viable explanation for soldiers to not take appropriate action when situations become morally or ethically corrupt? It certainly appears so. After all, even though soldiers are members of a profession, they are still human. Soldiers could be as motivated as anyone else to avoid embarrassment (audience inhibition) and may be just as likely to adopt behaviours that are similar to others when faced with ambiguous situations. Diffusion of responsibility could be even more pronounced in cohesive military units. Militaries train and function primarily as teams, so successes and defeats are a shared experience. Furthermore, given the hierarchical nature of military organizations, there is a clear establishment of responsibility. If someone senior is present, and not taking action, it is less likely that someone lower in the hierarchy will be willing to intervene. Confusion of responsibility may also play a part as soldiers try to distance themselves from the misconduct of unit mates for fear of being implicated in the wrongdoing. Although these factors may be used to explain why interventions do not take place, soldiers have a subsequent opportunity to make things right by reporting misdeeds to proper authorities. Regrettably, numerous cases have documented failures in this regard as well.

Whistle-blowers

Whistleblowing has been defined as the 'disclosure by organizational members of illegal, immoral, or illegitimate organizational acts or omissions to parties who can correct the wrongdoing' (Paul and Townsend, 1996: 150). A soldier's professional obligation to report the wrongdoings of one or more comrades would certainly fall within the auspices of this definition. There are two types of whistleblowing: internal and external. Internal whistle-blowers use avenues within the organization to report their wrongdoing while external whistle-blowers report the misconduct to authorities or others outside the organization (e.g. media). Lindblom (2007) refers to the moral dilemma of whistleblowing as stemming from a conflict between organizational loyalty and the right to speak out against wrongdoing. In the case of soldiers, loyalty to comrades will also be a deciding factor. Studies investigating whistleblowing have typically focused on the characteristics of the whistle-blower, the nature of the wrongdoing, aspects related to the perpetrator, and relevant contextual variables. In their meta-analysis, Mesmer-Magnus and Viswesvaran (2005) found that ethical judgement was only moderately related to whistleblowing intentions and completely unrelated to actual whistleblowing. In other words, there is evidently more to whistleblowing than simply knowing the right thing to do.

The failure to report misconduct, despite being aware of the moral require-ment to do so, is not surprising if we consider that whistleblowing has been asso-ciated with some extremely negative consequences. Jensen (1987: 325) reports that the punishments suffered by whistle-blowers 'have included loss of job, loss of financial security, loss of mutual respect from peers and superiors, loss of friendships, much psychological stress, general social ostracism, and even losing the confidence of one's family and loved ones'. In fact, retaliation against whistle-blowers by co-workers and organizations has been well documented (Dworkin and Near 1997; Bucka and Kleiner 2001; Rehg *et al.* 2008; Yeargain and Kessler 2010). Given the potential disruptions to work in particular, and life in general, it is not astonishing that people will fail to find the necessary moral courage to blow the whistle. Yet, whistle-blowers exist. Thus, one can only wonder, is there something inherently different about a whistle-blower?

In fact, those who demonstrate the required fortitude to become whistle-blowers do tend to share some common characteristics. According to Paul and Townsend (1996), whistle-blowers are more likely to be well educated, have high status, embrace egalitarian value systems, exhibit a strong professional ori-entation, and possess a high tolerance for rejection.

A quick review of these characteristics suggests that military members should be more likely than their civilian counterparts to report wrongdoings. After all, military members tend to be well educated, enjoy a position of relatively high status in society, should exhibit egalitarian values, and are recognized as profes-sionals. However, there is a possibility that the conditions necessary for a military group to function cohesively and effectively are not conducive to whistleblowing. According to Yeargain and Kessler (2010), Specialist (SPC) Joseph Darby's decision to report the violations taking place at Abu Ghraib has been devastating. He cannot return to his hometown, where his family lives, because the townsfolk regard him:

> as the traitor, not the seven who violated the code of military conduct. They are the heroes who can return home after they get out of prison. But Joe Darby who testified for the prosecution at their trials cannot. He is the one under witness protection.
>
> (Yeargain and Kessler 2010: 91)

Furthermore, even though he wished to remain anonymous, 'Secretary of Defense Donald Rumsfeld publicly named SPC Darby as the whistleblower during a U.S. congressional hearing televised worldwide by cable news net-works' (Brown 2008: 58).

According to Near *et al.* (2004), the nature of the wrongdoing itself can have an impact on the likelihood of whistleblowing. In their study, rates of whistle-blowing ranged from a low for waste and discrimination (17 per cent), to increased levels for sexual harassment (40 per cent) and mismanagement (43 per cent), and a high for legal violations (53 per cent). One would think that the nature of the abuse at Abu Ghraib would stimulate an abundance of reporting, yet only Joe Darby, out of the approximately 200 members present, exhibited the moral courage to blow the whistle. And the consequences for him were dire. This is in direct opposition to impressions offered by researchers like Sekerka *et al.* (2009: 567) that 'the military is a highly regulated organization with an emphasis on control to affect ethical behavior, it is characterized by a prevention orientation that is designed to curtail unethical practices'.

When considering the individual and situational characteristics associated with whistle blowing, is it fair to expect military members to mirror the behaviours of those working in civilian settings? Perhaps not. From the moment they don uniforms, military members are indoctrinated into a team environment where individuals seldom have the opportunity to work independently. As acknowledged by Sekerka *et al.* (2009), behaviours within the military are highly regulated and control is emphasized. This control is exercised over most behaviours and is not limited to ethical and moral activities. When members violate any expected norms of behaviour the consequences include formal and informal penalties that go far beyond anything that could be anticipated by their civilian counterparts. Within the military, ethical and moral violations will often take place in situations where serious injuries and death cannot be ignored as possible outcomes. Although whistleblowing may be desirable, the impact on cohesion, morale, teamwork, and trust can translate into decrements in performance and shortcomings in effectiveness. In short, the military environment may possess characteristics that make simple comparisons to civilian whistleblowing unrealistic.

Military professionalism

Over the past half-century many Western nations have transformed their militaries from large conscript forces to smaller forces of volunteers who are better trained and more motivated towards military service. This professionalization of military forces has two implications for our discussion of bystanders and whistleblowers. First, although soldiers in earlier times might have been forgiven for not displaying much in the way of personal discipline or professional behaviour – they were conscripts after all – this is no longer the case. Nowadays citizens have high expectations of their men and women in uniform and the media is unforgiving when it comes to the transgressions of military personnel. Second, military forces are smaller and leaders throughout the chain of command fewer in number than before, so junior soldiers are now expected to exercise more self-discipline and greater levels of professionalism than in an earlier age. Fifty years ago, Huntington stated that the military profession included only commissioned officers, and officers from the operational occupations at that (Huntington 1957/1985: 11). But views on the military profession have evolved since then to the point that when Canadian Forces professional doctrine was rewritten in the early 2000s, it was decided that everyone in uniform, regardless of rank or occupation, was to be considered part of the Canadian military profession (Canada 2003: 11).

So where does standing by while another member of your profession commits misdeeds fit with current notions of the military profession? And where does the failure to report the misdeeds of colleagues fit in? The simple answer is that these types of behaviour do not correspond with any concept of military professionalism. Hartle (1987) emphasizes that there is a need for military actions to be consistent with professional military ethics, while ensuring the protection of human rights, national interests, and the well-being of anyone involved. In other words, he envisions a type of military professionalism that would dictate intervention to save from harm those being abused and encourage the reporting of such misconduct to protect against future sufferings. For those who serve in uniform and, by definition, have dedicated their lives to protecting others, the failure to intervene when faced with the misconduct of colleagues, can only be described as an absence of professionalism. One would like to believe that highly trained soldiers would have gained some immunity to the behaviours associated with the bystander effect. There should be little doubt in any soldier's mind that there are no circumstances that can possibly justify torture, physical abuse, rape and murder. Thus, the ambiguity associated with social influence cannot be granted credence as a viable explanation. Similarly, it is almost inconceivable that a motivation to avoid embarrassment (audience inhibition) should come into play when soldiers are witnessing atrocities. Finally, the concept of diffusion of responsibility should not be counted as an acceptable factor. Professional soldiers are well aware of their responsibility to protect those who cannot protect themselves. Standing by, while others are abused in any way, is not a diffusion of responsibility, it is an abnegation of responsibility.

In 1988, the Nobel Peace Prize was awarded to the United Nations' Peacekeeping forces, acknowledging the contribution of current and former peacekeepers from around the world. Many Canadians welcomed this recognition with pride, for the Canadian Forces had earned an international reputation as a contributor to peacekeeping in many of the world's hot spots. Yet, a mere five years later, in March 1993, Canadians were embarrassed to learn that some soldiers of the Canadian Airborne Regiment, while deployed in Somalia, tortured and beat to death a Somali teenager, Shidane Arone. Understanding how professional Canadian soldiers could commit such a crime was difficult. According to Canadian Army Colonel and historian, Bernd Horn:

> What made this tragedy even harder to understand is the fact that throughout the beating, numerous soldiers, senior NCOs and officers either heard the cries or actually dropped by the bunker and witnessed the beating in progress; yet, no one stopped it until it was too late.
>
> (Horn 2001: 197)

A prominent Canadian historian suggested later that the fact that many heard the screams and did nothing indicated 'either a complete collapse of discipline or, just as likely, fear of their own soldiers among the officers' (Granatstein 2004: 155). Readers interested in a more in-depth analysis of how the bystander effect contributed to inaction by observers during this incident are encouraged to read Shorey (2000). He highlights the factors that led to the bystander effect in Somalia and concludes that a greater 'awareness of the potential damaging influence of such forces as the bystander effect and tacit authorization of misconduct seems a useful starting point to countering their development' (Shorey 2000: 27).

Whatever their reasons, the soldiers who turned a blind eye while a teenager was tortured to death are guilty of professional misconduct. Words like duty, integrity, ethics, and morality lose their meaning when they fail to influence soldiers to do the right thing. As indicated earlier, standing by while atrocities are being committed by fellow soldiers is inconsistent with any notion of military professionalism and duty. Adherence to the tenets of military professionalism requires personnel to intervene when witness to comrade misconduct. It is also important to highlight that the soldiers who were present to observe inappropriate behaviours without intervening, and then subsequently failed to report the misdeeds, were making a conscious choice to not get involved. These professional failures were not accidents, they were the consequence of a decision-making process.

Decision making

Remaining passive while unit colleagues engage in wrongdoing might look like the bystander or would-be whistle-blower is not doing much, but there is actually plenty of decision making underway, although much of it may be faulty. We will show in this section how decision-making errors and limits in moral competence can lead bystanders and potential whistle-blowers to neglect their professional responsibilities.

Decision-making errors

When reacting to the challenges of everyday problems, individuals employ two types of decision making. The first is called System 1 thinking by some scientists (Kahneman 2011: 20–30) and consists of intuitive, unconscious, rapid, automatic and effortless decision making. System 2 thinking on the other hand is rational, conscious, slower than System 1, controlled and effortful. We employ both styles of decision making in our daily life, but make greater use of System 1 than most of us realize, particularly in decisions involving morality. Both systems are fraught with errors and biases, and the list is too long to cover here, but we highlight five, which are particularly relevant to bystander and whistle-blower decision making.

Obedience to authority

Military personnel are inclined to defer to those in positions of authority, because that is what is taught in military training and emphasized in military culture.

Indeed, the success or failure of a military mission is dependent on an environment that reinforces 'trust in authority, willing obedience, and comradeship' (Aronovich 2001: 18). While this is beneficial for military discipline and effectiveness, it can contribute to faulty decision making in morally ambiguous situations or on those rare occasions when leaders actually direct subordinates to commit morally questionable acts. We should add that soldiers have been known to commit immoral acts in response to implicit suggestions from leaders (Kelman and Hamilton 1989). Leaders at all levels of the military chain of command should recognize that many soldiers have a natural tendency to comply with leaders and therefore may not always exercise sound critical thinking when directed by leaders, or even when speculating on what leaders might expect from them in uncertain situations.

Social proof

Social proof refers to the tendency of individuals to look to others for cues on how to act, particularly in ambiguous circumstances or in novel situations they haven't experienced before. The relevance to our current discussion is that 'people are more likely to undertake unethical actions in the workplace and elsewhere if peers are engaging in similar behaviour. And they are less likely to blow the whistle on unethical activity when peers seem to accept it' (Prentice 2004: 58).

Peer pressure

The military encourages conformity and social cohesion because they contribute to the operational effectiveness of military units. Unit affiliation is an important part of every soldier's personal identity, so it should come as no surprise that peer pressure can be a powerful force in the military. In a study of misconduct at the United States Naval Academy, Pershing (2002) found that loyalty to peers was stronger than loyalty to the organization. According to her, friendship is a key:

> factor in deciding how to react to alleged honor violations. It turned out to be the most important factor. All 40 graduates cited 'friendship or peer loyalty' as the first consideration in deciding whether to report a midshipman for occupational misconduct. That is, norms about peer loyalty typically translate into a 'code of silence' among friends when it comes to reporting honor violations.
>
> (Pershing 2002: 164)

Thus, peer loyalty could have a strong influence for both non-intervention and reporting. It could also be that military members have exaggerated tolerance levels for things like detainee abuse. In his study of veterans' acceptance of war zone detainee abuse, Holmes (2007) found surprisingly high levels of tolerance for abuse:

Veterans' zero tolerance for scenario-based depictions of detainee abuse that abrogated numerous Geneva Convention articles was low. Only little more than half of the participants believed sodomy with a broomstick, for example, was completely unacceptable even in a detainee about which nothing was known.

(Holmes 2007: 177)

In ethically-charged situations, soldiers will look for guidance from their leaders, both formal and informal, and in their absence, will turn to their peers. This will not lead to problems in units where there is a strong ethical climate, but it may in units where the ethical climate is weak, or where there are ethically-challenged leaders, or in demanding situations where leadership is absent.

Self-serving bias

According to social psychologists Carol Tavris and Eliot Aronson, 'all of us share the impulse to justify ourselves and avoid taking responsibility for any actions that turn out to be harmful, immoral or stupid' (Tavris and Aronson 2007: 2). As a result, we engage in self-deception and self-justification to protect our sense of self-worth. We can also expect some bystanders and would-be whistle-blowers to develop fanciful explanations as to why they didn't take professional action, because 'not only does the self-serving bias unconsciously affect the information that people seek out, causing them to search for confirming rather than disconfirming evidence, it also affects how they process that information' (Prentice 2004: 61).

Cognitive dissonance

Cognitive dissonance refers to the uncomfortable feelings we get when our thoughts, attitudes, or perceptions don't match our actions. The way we typically eliminate this dissonance is by either changing our actions or changing our attitudes about the actions we took or didn't take. For example, a soldier who witnessed another soldier abuse a non-combatant may try to reduce the dissonance associated with the fact that he didn't stop his peer from the wrongdoing by reframing the wrongdoing as legitimate behaviour, perhaps by convincing himself that the non-combatant deserved the abuse, or that the mission required aggressive action.

Moral competence

Potential bystanders and whistle-blowers may be limited in four critical moral competencies: moral perception, moral judgement, moral motivation, and moral implementation.[2] Moral perception refers to the decision maker's awareness of ethical issues in general and sensitivity to the ethical ramifications of the current situation in particular. Judgement refers to the reasoning capacity of the decision maker and his or her ability to evaluate the moral elements of the situation (i.e. potential consequences, moral obligations and virtues) for deciding what the proper moral response should be. Motivation refers to the extent to which the decision maker is guided by moral motives rather than pragmatic or self-serving motives. Implementation refers to the decision maker's capacity to convert his or her perception, judgement and motivation into a satisfactory professional response. These four components can work independently and in combinations to influence the actions of bystanders and whistle-blowers, so there are many ways in which shortcomings in moral competencies can derail ethical behaviour.

Let us consider these elements in the context of bystander decision making, which for ease of analysis we have reduced to three basic scenarios:

In Scenario 1, a soldier sees a unit mate do something wrong, recognizes the action as wrongdoing and intervenes. This scenario reflects proper professional reaction and we will say no more on it.

In Scenario 2 a soldier sees a unit mate do something wrong, but doesn't recognize the wrongdoing as inappropriate, and takes no action. This is an example of a failure in perception because the soldier was unable to correctly recognize the misbehaviour as a professional violation. It is important to note here that a failure in perception might explain a soldier's inaction, but doesn't excuse the inaction. Perhaps the bystander did not know what qualifies as acceptable behaviour in this circumstance or perhaps he/she had not anticipated that a colleague might behave in such a way. Either way, he/she was not prepared for this situation. Inaction of this sort can happen in situations where the wrongdoing occurs so quickly the bystander did not have time to perceive it fully or in cases where the misdeed is trivial, subtle or ambiguous. Sometimes a soldier's desire to get along with unit comrades can cloud his or her perceptions of what is right in such instances. Training can help with both types of failure and we will offer some suggestions later.

In Scenario 3, a soldier witnesses a transgression and correctly recognizes it as misconduct, but does not intervene. In this case, the soldier had sufficient ethical awareness to perceive the observed action as wrongdoing, but chose not to intervene, possibly as a result of a failure in either judgement or motivation. Even though the observer recognized the ethical demands of the situation, he/she may have incorrectly judged that no response was needed. This can happen for a variety of reasons which we proposed above in our coverage of decision-making errors. Failures of motivation are common and may also be magnified by decision-making biases. In the case of motivational failure, the observer recognizes the moral dimensions of the situation, realizes that he or she should take action, but doesn't because the cost of taking action will be too severe (e.g. disdain of colleagues, threat of retaliation) and he/she can't find the moral courage needed to overcome these self-serving motives. We will show later in the chapter that education and training can help improve moral judgement, but it is less clear that moral motivation can be enhanced with instruction, although we have some suggestions on this.

We will now consider whistle-blower decision making as it relates to the moral competencies described above. We suggest three basic decision-making scenarios:

In Scenario 4, a would-be whistleblower is aware of some wrongdoing – perhaps he witnessed it or only heard about it. In this scenario, the soldier correctly identified the wrongdoing and reported it, which is the proper professional reaction, so we will say no more on it.

In Scenario 5, the soldier doesn't recognize the wrongdoing as improper and does not report it. This case represents a failure in perception because the soldier was unable to correctly recognize the transgression as misconduct that should be reported. It is also possible that the soldier recognized that the behaviour was improper and should be reported, but didn't because he didn't know how to report it. Fortunately, failures in moral perception from a lack of moral awareness can be corrected with additional training.

In Scenario 6, a soldier is aware of some misconduct which he correctly identifies as wrongdoing, but he decides not to report it. This could be a failure of moral judgement or moral motivation. On the judgement side, the soldier may have reasoned that it is not his or her role to report the violation, or that the misconduct was not serious enough to warrant reporting. The failure might also be due to lack of moral motivation (moral courage), for as we mentioned above the personal costs of whistleblowing can be high, particularly in cohesive organizations. There is also the possibility, albeit remote, that the soldier did not report the offence because he/she did not know how to report it or to whom it should be reported. If this was the case, it also reflects a failure in moral motivation to some extent because the soldier should have asked others for information on how to report the misbehaviour. As previously mentioned, shortcomings in awareness and judgement can be remedied with additional ethics instruction, but deficiencies in moral motivation are more difficult to correct.

Our discussion of bystander and whistle-blower decision making in this section has been somewhat hypothetical, but the unfortunate reality is that there is very little research on military bystanders and whistle-blowers and even less on the decision-making processes they employ. At the time of writing this chapter, we were aware of only two studies which had examined these issues in the military milieu. The first was the work of a U.S. Mental Health Advisory Team which asked soldiers and marines serving in Operation Iraqi Freedom if they would report a unit member for committing ethical violations on the battlefield. Overall, the number of respondents who agreed that they would report a unit comrade for injuring or killing an innocent noncombatant seems low (55 per cent of soldier respondents and 40 per cent of marine respondents) although this study does not appear to have been replicated either in the U.S. or elsewhere (MHAT IV), so we cannot be certain that the numbers of those who say they would report this transgression are indeed low or not. The second study is based on surveys of

Canadian soldiers returning from the war in Afghanistan which asked them the extent to which they agreed with the statement, 'I would report breaches to the Canadian Forces Code of Conduct and Law of Armed Conflict even if it meant I would be subject to retaliation from fellow soldiers'. Recorded in a Canadian Forces research report by Messervey (forthcoming), the findings from five Canadian task forces were fairly consistent with average agreement rates of 40 per cent for junior non-commissioned members, 70 per cent for senior NCOs, 80 per cent for junior officers and 90 per cent for senior officers. The attitudes towards reporting increase positively with rank but, alarmingly, more than half the junior ranks group (private – master corporal) indicated they would not report their comrades. Together, these two studies suggest that there is a bystander problem in the military, particularly at the lower ranks, where problems can worsen if unit leaders never learn about them.

To the extent that the military institution can solve the bystander/whistle-blower problem, the solution will likely be found in one or both of the two components of the personnel production system which the military employs to ensure its members possess the knowledge, skill, and abilities they need to perform successfully – the training system and the selection system. For selection to be the preferred option to the bystander/whistle-blower problem, moral functioning must be a stable characteristic. Training will be the more efficient solution if moral competence can be altered. To the extent that a trait can be developed, it is not a suitable marker for selection. Military training is used to develop skills that are critical to success on the battlefield like marksmanship and navigation, but these are also skills that few recruits possess on joining the military. Fortunately, they can be developed. In fact, most nations have extensive military training systems to provide their soldiers and officers with the knowledge and skills they need to perform their duties. There is also substantial research evidence that ethical functioning can be enhanced with education and training (there will be more on this later in the chapter).

Selection

In order for selection to be the best option for acquiring ethical soldiers, several conditions must be met. First, we must be able to measure moral functioning accurately and second, performance on the selection test must be related to important aspects of subsequent, on-the-job performance. There is an emerging field of honesty and integrity testing that shows promise for moral selection. In fact, many Western countries already conduct background checks on recruits by submitting their names to law enforcement databases to identify those with criminal convictions so they can be screened out as unsuitable candidates. Tests might be helpful as well. Many civilian companies employ integrity tests to screen new employees, mostly applicants for low-level, non-managerial jobs, a category of workers very similar to military recruits. Studies show that many of these tests can measure the attributes of honesty and integrity reliably (Berry *et al.* 2010: 271–301) and a number of these tests are able to predict subsequent, on-the-job, counterproductive behaviours like theft (Wanek 1999: 183–195) and substance abuse (Ones *et al*. 1993: 679–703) making them suitable as screening tests.

There are several problems with integrity tests, however. First, they are relatively transparent, so dishonest individuals who are bright enough can fake their responses to pass the tests. Second, integrity tests are like all selection tests in that they have a margin of error which means that some of the people who fail the test might actually have gone on to perform successfully on the job. Unfortunately, there is no way of knowing how many people would be incorrectly screened out in this way. Third, a large problem with military selection tests is simply the length of time it takes to administer each test to recruits. The enrolment processes in most Western militaries already demands a significant amount of time for testing, medical examinations, and additional administrative requirements, so authorities are reluctant to add other elements to the selection process unless the benefits are substantial.

In the end, selection can be used to improve soldier morality. However, if we accept that one of the roles of a nation's military is to provide social mobility opportunities for young men and women, perhaps the moral option is to rely mostly on training solutions and employ selection measures only to screen out the most unethical recruits.

Training

It is broadly accepted that if individuals are deficient in a particular area, their performance can be improved with instruction and most military forces have extensive training programs in place to ensure their personnel learn the skills they need to perform successfully. The militaries of the Western world assign plenty of resources to ethics instruction, but there is little evidence that these efforts have much effect. We know of only one published study which has attempted to show that ethics training works in the military (Warner *et al*. 2011) and we will speak more on it later. Studies of ethics instruction effectiveness outside the military have produced mixed results. In fact, several authors have explicitly stated that ethics instruction does not affect student behaviour (Orwin 2009; Dean and Beggs 2006). One of these, a telephone survey of 27 faculty who teach business ethics in two U.S. universities found that business ethics courses had 'little or no impact on students' ethical behaviours' (Dean and Beggs 2006: 40).

However, criticizing ethics instruction because it does not lead to visible changes in ethical behaviour reflects a narrow view of moral functioning. Moral behaviour consists of more than observable actions. It also includes less visible behaviours such as perceptions, decision making and motivation as described by James Rest and his colleagues in their four-component model of morality (Rest 1986; Narvaez and Rest 1995). Part of the problem with training effectiveness studies lies in measuring ethical behaviour, because objective measures of what is essentially a dichotomous variable (the individual acted ethically or did not act ethically) are very hard to obtain. The alternative for many researchers is to gather self-reports through surveys or interviews, though self-reports can be unreliable.

When we view moral functioning as a multidimensional concept (as we described earlier) – consisting of moral perception/sensitivity, moral judgement, moral motivation and moral implementation/action – studies are able to gain a better appreciation of the full range of moral behaviour and the complex relations between instruction and ethical competence. For example, a recent review of 25 studies involving a total of 6,791 business ethics students found that the correlations between instruction and ethical outcomes were different for moral judgement, moral sensitivity (i.e. perception) and moral action (Waples *et al.* 2008: 139). The association between ethics education and all outcomes was minimal, but it was strongest in the case of the instruction–moral judgement linkage and weakest between instruction and ethical behaviours.

Studies typically find that ethics education is more predictive of ethical sensitivity and ethical judgement than other ethical outcomes, but the relationships are usually not strong. Unfortunately, there are no published accounts of ethics instruction influencing moral motivation. An early researcher in the field of ethical sensitivity, Muriel Bebeau, reported that ethical sensitivity could be improved with instruction, but not always (Bebeau 2002: 284). Waples and colleagues found that ethics instruction was related to ethical sensitivity, albeit moderately (Waples *et al.* 2008: 139). Similarly, when the phone interview study of business ethics faculty described earlier turned its attention to moral perceptions, the instructors reported that their courses were able to alter students' moral sensitivity if not their actual moral behaviour (Dean and Beggs 2006: 34).

More research has been conducted on the linkage between education and moral judgement than with any other moral component. In fact, there is strong evidence to show that education leads to improvements in moral judgment when measured by instruments like the Defining Issues Test and Kohlberg's Moral Judgment Interview (Colby 2008: 400). Pascarella and Terenzini (1991; 2005) conducted several reviews of this research and concluded that 'college experience itself has a unique positive influence on increases in principled moral reasoning' beyond that which would be expected simply from maturation (347). In their most recent review, Pascarella and Terenzini (2005) noted that students generally advanced from Kohlberg's Level II conventional moral reasoning to Level III post-conventional reasoning (Kohlberg 1976 34–35) as they progressed through their university studies. Other research has shown that an undergraduate liberal arts education will typically improve one's moral judgement, but more specialized and technical postgraduate programs like medicine, dentistry and law do not, unless there is an ethics component in the curriculum.

The research on ethics instruction effectiveness described above was conducted mostly on university-level educational programs, but most military ethics instruction is properly classified as training. The instruction that junior personnel receive in their military units is typically of short duration and developed specifically with operational roles in mind, so it is more like training than education.

Unfortunately, there is not a lot of research on the effectiveness of ethics training and the results of what is available are not promising. For example, a review of ethics interventions by Schlaefli *et al.* (1985: 342–347) found that programs of less than three weeks' duration did not lead to increases in moral judgement.

Some studies suggest that ethics education may be more effective than ethics training, but there is not much work on this issue. More research is required before firm conclusions can be drawn on the relative effectiveness of ethics education and training. In this chapter, we use the broader term 'ethics instruction', which refers to both educational and training interventions.

However, there is one encouraging study on the effectiveness of ethics training and it was conducted in the military. The training consisted of leader-led 'structured discussions of specific movie vignettes involving ethical dilemmas' in a U.S. Army Stryker Brigade (Warner *et al.* 2011: 917). Leaders throughout the chain of command of the brigade led these discussions with their direct-report subordinates (i.e. subordinates who reported directly to them). As these subordinates completed the training with their commander, they then led structured discussions with their direct subordinates and so on down the chain of command. Of the roughly 3,500 personnel who took the training, 397 agreed to complete a post-training survey. Comparison of the pre-training and post-training surveys found that:

> decreased rates of unethical conduct were noted in all categories after training, with significant reductions in reports of insulting and cursing at non-combatants, unnecessary damage or destruction of private property, or witnessing mistreatment by a fellow unit member. Significant reductions were seen in all categories of attitudes related to reporting of a fellow soldier for ethical misconduct ... significant increases were seen in the willingness to report ethical violations
>
> (Warner *et al.* 2011: 920–921)

In summary, research shows that university-level ethics courses can lead to improvements in ethical sensitivity and judgement. Other research shows that increasing ethical sensitivity and judgement can lead to increased ethical behaviour. Unfortunately, there is little research showing how to raise levels of moral motivation and the impact of ethics training of shorter duration on moral functioning is not yet clear, although the Warner *et al.* (2011) study is promising.

Recommendations

In this chapter, we have suggested that some soldiers may be reluctant to intervene when their colleagues engage in misconduct or to report the misconduct to proper authorities. Admittedly, the evidence for these problems is slim, because there have been very few studies on these issues and soldiers are probably reluctant to talk about these matters outside their peer groups. Clearly, more research is needed in this area.

Given the research evidence that ethical awareness and judgement can be improved with instruction, we recommend training interventions directed at two levels – organizational culture and individual professional identity. We know from a long line of research that the behaviour of individuals is heavily influenced by situational factors (Zimbardo 2007), so the optimal way to reduce bystander apathy and increase responsible internal whistleblowing is to ensure that both actions are seen by soldiers as consistent with the norms and culture of their military unit. Therefore, leaders at all levels of the military hierarchy must be taught to show subordinates that standing by while bad things happen will not be tolerated in the unit and that soldiers have a responsibility to tell unit leaders about any misconduct which occurs. As for individual responsibility, everyone in the chain of command regardless of rank or position must be taught that they have a professional duty to intervene when unit members transgress and to report transgressions to proper authorities. We recommend that training be directed explicitly at the four moral components discussed earlier in the chapter and offer some thoughts now on each.

Moral awareness, perception and sensitivity

Soldiers will not intervene or report wrongdoing if they don't realize that the action they just witnessed is actually misconduct. Therefore, training has to begin with what is acceptable behaviour, what isn't acceptable, and the types of misbehaviour that soldiers will likely face in the foreseeable future. This element of training has to be updated regularly because some (not all) ethical challenges are situation-specific, meaning that they will vary from mission to mission and across military occupations. There will be unique ethical challenges, for example, in urban conflict that are not as prominent in conventional operations. Similarly, some of the ethical dilemmas confronting soldiers who handle detainees will differ from the problems challenging medical technicians working in a field hospital. Also, it is not good enough to offer this sort of instruction during basic training and then leave it; it has to be recurring and revised regularly.

Moral judgement

Like most people, soldiers can distinguish right from wrong if given enough time and no distractions. But in those situations where emotions are high and social influences are dominant, thinking can become distorted and behaviour can bend to prevailing forces, both good and bad. As a result, soldiers need to be taught how to conduct an ethical analysis, to overlearn the process in fact, just like they overlearn battle drills, so they will be able to make decisions quickly and correctly when put to the test in stressful situations. Along with rapid ethical problem solving, soldiers should be taught about common decision-making errors and biases, and given tips on how to manage these threats to ethical judgement.

Moral motivation

One of the major obstacles that soldiers will have in solving bystander/whistle-blower dilemmas is subordinating their personal motives to higher-order professional obligations. As mentioned above, most individuals will know what is right in most situations, it's just that they can't always will themselves to do the right thing, and then they invoke convoluted rationalizations to reassure themselves they did nothing wrong. One way to strengthen moral motives is with training sessions on professional identity, in particular the responsibilities and motives that define what it means to be a professional soldier in this military force, this unit, etc. The aim here is to get soldiers to commit to professional expectations and to align their personal and social motives with the professional norms of intervening and reporting when the situation demands.

Moral action and implementation

Soldiers should be taught how to intervene when a peer or subordinate is doing something wrong or how to speak up when a superior officer is doing something wrong. They should also be taught how to report transgressions, because junior personnel in particular may not know what to do when one of their leaders is involved in wrongdoing. Active bystander instruction can include role playing exercises and tips on how to intervene. Avenues of reporting within the chain of command should be encouraged because internal whistleblowing supports the development of a healthy organizational culture, but soldiers should also be taught about external reporting mechanisms (ombudsman, hot lines, etc.) in case they are thwarted by unresponsive leaders.

As a final note, we recommend that ethics instruction of this sort be embedded within professional training, and perhaps even called professional training or decision making rather than ethics training, because the term 'ethics' can provoke resistance from some quarters and diminish the impact of the instruction.

Notes

1 In this chapter, the term soldier can refer to anyone in uniform, whether they be male or female, or belong to the navy, air force or army.
2 Adapted from Rest's four component model of morality (Rest 1986; Narvaez and Rest 1995).

Bibliography

Aronovich, H. (2001) 'Good soldiers, a traditional approach', *Journal of Applied Philosophy*, 18(1), 13–23.
Bebeau, M.J. (2002) 'The defining issues test and the four component model: Contributions to professional education', *Journal of Moral Education*, 31(3), 271–295.
Berry, C.M., Sackett, P.R. and Wieman, S. (2007) 'A review of recent developments in integrity test research', *Personnel Psychology*, 60(2), 271–301.

Brown, W.E. (2008) 'Whistleblower protection for military members', *The Army Lawyer*, *DA PAM 27–50–427*, 58–65.

Bucka, D. and Kleiner, B.H. (2001) 'Whistleblowing in the aerospace and defence industries', *Managerial Law*, 43(1–2), 50–56.

Cacioppo, J.T., Petty, R.E. and Losch, M.E. (1986) 'Attributions of responsibility for helping and doing harm: Evidence for confusion of responsibility', *Journal of Personality and Social Psychology*, 50(1), 100–105.

Canada (2003) *Duty with Honour: The Profession of Arms in Canada*. Canadian Defence Academy – Canadian Forces Leadership Institute, Kingston, Ontario, online at http://publications.gc.ca/collections/collection_2011/dn-nd/D2-150-2003-1-eng.pdf.

Colby, A. (2008) 'Fostering the Moral and Civic Development of College Students', in Larry P. Nucci and Darcia Narvaez (ed.) *Handbook of Moral and Character Education*, New York and London: Routledge.

Cook, K. (2014) *Kitty Genovese: The Murder, the Bystanders, the Crime that Changed America*, New York: W.W. Norton.

Darley, J.M. and Latane, M. (1968) 'Bystander Intervention in Emergencies: Diffusion of Responsibility', *Journal of Personality and Social Psychology*, 8, 377–383.

Dean, K.L. and Beggs, J.M. (2006) 'University Professors and Teaching Ethics: Conceptualizations and Expectations, *Journal of Management Education*, 30(1), 15–43.

Dworkin, T.M. and Near, J.P. (1997) 'A Better Statutory Approach to Whistle-Blowing', *Business Ethics Quarterly*, 7(1), 1–16.

Frederick, J. (2010) *Black Hearts: One Platoon's Descent into Madness in Iraq's Triangle of Death*, New York: Harmony Books.

Friscolanti, M. (2010) A Soldier's Choice. *Maclean's Magazine*, 24 May 2010, 20–25.

Friscolanti, M. and Geddes, J. (2010) 'A Stern Message About Battlefield Ethics and the "Soldier's Pact"', *Maclean's Magazine*, 2 August 2010, 28–30.

Gage, W. (2011) *The Report of the Baha Mousa Inquiry – Volume 1*, London: The Stationery Office.

Granatstein, J.L. (2004) *Who Killed the Canadian Military?* Toronto, Ontario: Harper Collins Publishers Ltd.

Hartle, A.E. (1987) A Military Ethic in An Age of Terror, *Parameters*, Summer, 68–75.

Holmes, W.C. (2007) 'Abuse of War Zone Detainees: Veterans' Perceptions of Acceptability', *Military Medicine*, 172(20), 175–181.

Horn, B. (2001) *Bastard Sons: An Examination of Canada's Airborne Experience 1942–1995*, St Catherines, Ontario: Vanwell Publishing Limited.

Huntington, S.P. (1957/1985) *The Soldier and the State: The Theory and Politics of Civil-Military Relations*, Cambridge: Belknap.

Jensen, J.V. (1987) 'Ethical Tension Points in Whistleblowing', *Journal of Business Ethics*, 6, 321–328.

Kahneman, D. (2011) *Thinking, Fast and Slow*, Doubleday Canada.

Kelman, H.C. and Hamilton, V.L. (1989) *Crimes of Obedience: Towards a Social Psychology of Authority and Responsibility*, New Haven and London: Yale University Press.

Kohlberg, L. (1976) 'Moral Stages and Moralization', in T. Lickona (ed.) *Moral Development and Behavior: Theory, Research, and Social Issues*, New York: Holt, Rinehart and Winston.

Latane, B. and Darley, J.M. (1968) 'Group Inhibition of Bystander Intervention in Emergencies', *Journal of Personality and Social Psychology*, 10(3), 215–221.

Latane, B. and Nida, S. (1981) 'Ten Years of Research on Group Size and Helping', *Psychological Bulletin*, 89(2), 308–324.

Lindblom, L. (2007) 'Dissolving the Moral Dilemma of Whistleblowing', *Journal of Business Ethics*, 76, 413–426.

Mental Health Advisory Team (MHAT) IV Operation Iraqi Freedom 05–07, Final Report, 17 November 2006, online at http://i.a.cnn.net/cnn/2007/images/05/04/mhat.iv.report.pdf.

Mesmer-Magnus, J.R. and Viswesvaran, C. (2005) 'Whistleblowing in Organizations: An Examination of Correlates of Whistleblowing Intentions, Actions, and Retaliation', *Journal of Business Ethics*, 62, 277–297.

Messervey, D. (Forthcoming) *What Drives Moral Attitudes and Behaviour?* Defence R&D Canada – Technical Report, DRDC TR.

Narvaez, D. and Rest, J.R. (1995) 'The Four Components of Acting Morally', in W.M. Kurtines and J.L. Gewertz (eds.) *Moral Development: An Introduction*, Boston, MA: Allyn and Bacon.

Near, J.P., Rehg, M.T., Van Scotter, J.R. and Miceli, M.P. (2004) Does Type of Wrongdoing Affect the Whistle-Blowing Process? *Business Ethics Quarterly*, 14(2), 219–242.

Ones, D.S., Viswesvaran, C. and Schmidt, F.L. (1993) Comprehensive Meta-Analysis of Integrity Test Validities: Findings and Implications for Personnel Selection and Theories of Performance, *Journal of Applied Psychology Monograph*, 78, 679–703.

Orwin, C. (2009) 'Can We Teach Ethics? When Pigs Fly', *Globe and Mail*, 6 November 2009, online at www.globecampus.ca/in-the-news/article/can-we-teach-ethics-when-pigs-fly/.

Pascarella, E.T. and Terenzini, P.T. (1991) *How College Affects Students: Findings and Insights from Twenty Years of Research*, San Francisco: Jossey-Bass.

Pascarella, E.T. and Terenzini, P.T. (2005) *How College Affects Students: A Third Decade of Research Vol. 2*, San Francisco: Jossey-Bass.

Paul, R.J. and Townsend, J.B. (1996) 'Don't Kill the Messenger! Whistle-Blowing in America – A Review and Recommendations', *Employee Responsibilities and Rights Journal*, 9(2), 149–161.

Pershing, J.L. (2002) 'Whom to Betray? Self-Regulation of Occupational Misconduct at the United States Naval Academy', *Deviant Behavior: An Interdisciplinary Journal*, 23, 149–175.

Prentice, R. (2004) 'Teaching Ethics, Heuristics, and Biases', *Journal of Business Ethics Education*, 1(1), 55–72.

Rehg, M.T., Miceli, M.P., Near, J.P. and Van Scotter, J.R. (2008) 'Antecedents and Outcomes of Retaliation Against Whistleblowers: Gender Differences and Power Relationships', *Organization Science*, 19(2), 221–240.

Rest, J. (1986) *Moral Development*, New York: Praeger Publishers.

Robinson, P., de Lee, N. and Carrick, D. (2008) *Ethics Education in the Military*, Aldershot, Hampshire, UK: Ashgate.

Schlaefli, A., Rest, J. and Thoma, S. (1985) 'Does Moral Education Improve Moral Judgment? A Meta-Analysis of Intervention Studies Using the Defining Issues Test', *Review of Educational Research*, 55(3), 342–347.

Sekerka, L.E., Bagozzi, R.P. and Charnigo, R. (2009) Facing Ethical Challenges in the Workplace: Conceptualizing and Measuring Professional Moral Courage, *Journal of Business Ethics*, 89, 565–579.

Shorey, G. (2000) Bystander Non-Intervention and The Somalia Incident, *Canadian Military Journal*, Winter 2000–2001, 19–28, online at www.journal.forces.gc.ca/vol1/no4/doc/19-28-eng.pdf.

Tavris, C. and Aronson, E. (2007) *Mistakes Were Made (But Not by Me): Why We Justify Foolish Beliefs, Bad Decisions, and Hurtful Acts*, Orlando, Florida: Harcourt.

Wanek, J.E. (1999) Integrity and Honesty Testing: What Do We Know and How Do We Use It? *International Review of Selection and Assessment* 7(4), 183–195.

Waples, E.P., Antes, A.L., Murphy, S.T., Connelly, S. and Mumford, M.D. (2008) A Meta-Analytic Investigation of Business Ethics Instruction', *Journal of Business Ethics*, 87, 133–151.

Warner, C.H., Appenzeller, G.N., Mobbs, A, Parker, J.R., Warner, C.M., Grieger, T. and Hoge, C.W. (2011) 'Effectiveness of Battlefield-Ethics Training During Combat Deployment: A Programme Assessment', *The Lancet*, 378, 915–924.

Yeargain, J.W. and Kessler, L.L. (2010) 'Organizational Hostility Toward Whistleblowers,' *Journal of Legal, Ethical and Regulatory Issues*, 13(1), 87–92.

Zimbardo, P. (2007) *The Lucifer Effect: Understanding How Good People Turn Evil*, New York: Random House.

5 Ethics at and after war
Challenging battlefields[1]

Stéphanie A.H. Bélanger and Michelle Moore

Introduction

When the Canadian Armed Forces' Law of Armed Conflict at the Operational and Tactical Levels[2] (LACOTL) was introduced in 2001, soldiers became legally responsible for behaving ethically when in combat. The Canadian Armed Forces (CAF) adjusted their indoctrination process to incorporate ethical training in an attempt to provide soldiers with resources to enable them to carry out their duties in accordance with these new requirements, in all circumstances. Issued under the authority of the Chief of Defence Staff, the LACOTL is a complementary publication to the Code of Conduct for CAF Personnel, superseding the Manual and Unit Guide on the Geneva Convention of 1949, issued in 1973 and 1990 respectively.[3] The Code of Conduct within the LACOTL, which every CAF member is expected to uphold, details the basic principles and spirit of the law as the minimum required standard expected during all Canadian military operations overseas.[4] This duty to maintain the expectation of public trust in the soldier's professional and ethical behaviour is identified as an operational imperative.

The CAF Chief of Land Staff issued its guidance on ethics in operations in the most recent publication on the code of conduct in 2009, entitled *Duty with Discernment*.[5] Through the creation of the 'Canadian Soldier's Code of Conduct' that can be found in Appendix 3 of *Duty with Discernment*, a military ethos was condensed, theorized, inscribed and then incorporated into all aspects of CAF training and operations. The Code of Conduct is divided into eleven rules intended to guide soldiers in behaving in accordance with the LACOTL.[6] The operational planning process is continually scrutinized under this Canadian code of ethics, making ethics *a legal requirement rather than an option*: 'Ethics are not, repeat not, an optional extra.'[7] Failing to behave ethically becomes 'a criminal act.'[8] The aim of embedding the ethical behaviour into a code of law is to force, through indoctrination, CAF personnel to adopt a pattern of behaviour that is always ethical and applicable to combat situations in land operations.

Interviews have been conducted with 75 combat-ready CAF personnel to address the ways in which the Code of Conduct, more precisely the one published in *Duty with Discernment*, is lived on an individual basis.[9] A discourse analysis allowed the measuring of the impact of indoctrination on the soldiers identities through the common locus and through analysis of the contradictions that were used in their description of their experience of war.[10] 'Discourse analysis' in the context of the study of testimonies[11] allowed identifying, interpreting, recognizing, and appreciating the impact of war on soldiers, more precisely, on soldiers 'identities', as opposed to 'identity'.[12] Discourse analysis recognizes the postmodern vision of language as being unstable throughout different testimonies and within the same interview. It demonstrates that the linguistic conflicts arising from interviewing soldiers can be seen as a dissociation between the personal identity of the soldier and the construction of the self as a soldier. It allows approaching what one may see as the monopoly of the military identity, not as a 'ravage of the self' (Dawes 2002: 178), but rather as a tool for survival that undergoes many transformations throughout the military experiences of soldiers. It also argues that the many phases of transitions while returning home allow the soldiers to question their learned identity, to adjust their conception of their ethical behaviour. The 'deconstruction' impact of the experience of war has allowed for the better understanding of the perception that military members have of themselves, of their roles, and of their identity.[13] In a world where all is fragmented, where ideologies have failed for being the leading proponent of atrocities, and where the truth is relative, the study of testimonies becomes the study of multiple identities: individual, social and national.

Soldiers rely on what they have learned during the indoctrination process, which would normally provide them with the basic understanding of what is expected of them. For instance, a 2009 survey by *My Say* revealed that 87.8 per cent of the soldiers who have participated in the survey reported that they were satisfied with their job.[14] Similarly, the analysis of the examples used by the 75 interviewed soldiers while describing their deployment experiences shows the strength of the indoctrination on discourse, if not on the soldiers' identities. A vast majority of military members expressed enjoying a harmonious relationship with their job; more than eight soldiers out of ten considered that they did their work according to the rules and requirements of their profession and in conformity with their training.

In other words, Canadian soldiers say unanimously that they are satisfied with their job and that it corresponded with what they were expecting. These expectations are first seeded during basic military training. The organisation of the training imposes a variety of new behavioural adjustments: new ways of thinking, of expressing oneself (and even of walking!), of offering marks of respect and of behaving in groups from the very start of their career, during the first few weeks of indoctrination:

> In the Army, combat skills and values are learned through a socialization process that begins when a soldier enters basic training, which aims to instil new attitudes, responses and loyalties in the recruit as it is teaching him, or her, new skills.
>
> (Winslow 1998: 352)

After this initial period of about a month, the recruit assimilates those mental and physical reflexes that are considered as essential to the survival of a soldier within a cohesive group, as well as for his future survival in a combat situation:

> Hum, well, huh, in times of extreme, huh, anxiety, no that's a bad word. Huh, in times of extreme, when you get shot at or whatever, you don't, huh, you always fall back on your training.
>
> (File 60)[15]

This training is not the simple execution of given orders via the chain of command, but the result of pro-active collective efforts. In the battlefield this depersonalisation, this systematic, automatic organisation of the rules, gives the members a sense of comfort in familiarity. For those in command and those composing the platoon, the mission becomes something logical and coherent, within which soldiers only have to apply the rules. Group cohesion becomes second nature and this ensures survival in a situation which is no longer simulated, where the environment is chaotic and controlled by a possibly merciless enemy with no respect for legal or ethical constraints on behaviour in wartime. What would be considered unnecessarily aggressive behaviour in any other circumstances in any other organisation, is a necessary tool for survival for soldiers, whether during their training or in deployment. Since the training is crucial in determining each of the decisions taken and each of the experienced emotions, it is difficult for an impartial observer to evaluate what, in the report between the organisation and the member, is an objective observation or a soldier's simple reflex. This organisational force can explain why, out of 75 interviews, the correlation between the words that relate to war and the words that reflect a positive experience are much stronger and more confidently and positively expressed when the soldiers, during interview, described and explained their experience in Afghanistan. The terms most frequently used by soldiers were the following: 'soldier' (123), 'job' (114), 'military' (101), 'Afghanistan' (87). The word 'peace' appeared infrequently (17 times across 75 testimonies), whereas the word 'war', for example, appeared 146 times: it was as if any other activity was considered as secondary:

> I'm not saying that it's a game, but you go to hockey practice so you can play the game just like we train for war so we can go overseas and deploy. So, it's good to get that experience in, to finally have that sense of accomplishment, to do something you joined the army to do. On the other hand, you're there so you can help the Afghans out of there.
>
> (File 1)

Going to war is essentially what soldiers train for and, taken in this perspective, the discourse seems transparent: the soldiers have a sense of accomplishment when they participate in a war mission. But beyond these obvious semantic correlations, a more in-depth discourse analysis shows many semantic transformations of the words used to designate the war experience. These semantic changes unveil discursive contradictions that surface in their discourse, showing the pressures and cracks within the training and reintegration process. It is suggested that these inappropriate comparisons reveal the areas of a soldier's experience where they were unable to effectively embody the organizational expectations needed to make sense of their own experience; where they were unable to link their experience to the organizational explanation of their role.

The soldier's ethos within the CAF is defined as Duty, Loyalty, Integrity, and Courage,[16] and encompasses this soldierly identity that is taught, adopted, and reinforced from indoctrination through to, and after, operations. The soldiers use each of these four ethical values across the collected testimonies to try to come to terms with their experiences. Some contradictions have become apparent when the soldiers are unable to reconcile their lived experiences with this learned ethos, and will be discussed below.

The soldier and his shield

Within the manual *Duty with Discernment* are images of the Canadian soldier, predominantly portrayed as an infantry man carrying a C7 rifle, a side arm, and dressed in Full Fighting Order (FFO). Rather than stalking his enemy with his weapon at the ready, this soldier walks with a neutral expression, his rifle down at his side and a visible radio suggesting his ability and willingness to communicate. This modern representation of the soldier effectively illustrates the mentality behind the ethical code of conduct. The CAF member *is expected to avoid the use of force except when it is absolutely necessary*, as outlined throughout the eleven specific points within the Code of Conduct. The soldier must be, and is meant to be perceived as, a protector rather than an aggressor, and this intent is reflected by his protective military gear and lowered weapon. But this image, though focused on the defensive posture of the soldier, also shows the protector's potential for lethal action if a threat is imminent. The use of protection to portray strength and force is not unique to the CAF or a modern innovation, but rather belongs to a long tradition in the Western world. In this respect, the FFO can be seen as the modern representation of the coat of arms, the bronze hat and the shield of ancient warriors, as described in detail in Hanson's *The Western Way of War*.[17]

The first Occidental representation of the warrior insisted on the shield as being the most important military asset. Achilles' shield, described over many verses in the eighteenth book of the *Iliad*, displays first the elements of antiquity: the earth, sky, sea, sun, and moon. It then details the splendid cities, some engraved with images of joyful weddings and others with admired judges resolving conflict using the principles of justice; some others simply portray agriculture through images of vines and flocks. As the shield served as the most important asset to a warrior, the illustrations on the shield encompass all of the basic elements of a civilisation's culture, as these are a source of pride and inspiration for the warrior: they represent everything that the warrior is protecting.

These images displayed on the shields of warriors signify an ordered society composed of free citizens who make the right decisions and behave according to a code of law that will praise the best among them, the heroes, the η΄ρως, the 'half-gods'.[18] The shield was the greatest gift that a god could offer to a human being, as it implied protection by the gods, but also inspired terror even though it was not an aggressive weapon. The focus in historical representations of Achilles' behaviour did not always include his anger: 'Sing, Goddess, sing of the rage of Achilles, son of Peleus' says the first verse. Though some specific cultures and historical episodes remained inspired by this demonstration of strength through aggressiveness,[19] most of the occidental history retained the 'protection forces' as their main warrior value. Protection was such a strong value at the peak of the later Roman Empire that it was used as a symbol of fidelity to the Emperor by many authors. For instance, Plutarch, in his *Parallel Lives*, more specifically the *Life of Caesar*, illustrates the impact the general had on his troops using three examples involving the importance of the shield, principally this one:

> Again, in Britain, when the enemy had fallen upon the foremost centurions, who had plunged into a watery marsh, a soldier, while Caesar in person was watching the battle, dashed into the midst of the fight, displayed many conspicuous deeds of daring, and rescued the centurions, after the Barbarians had been routed. Then he himself, making his way with difficulty after all the rest, plunged into the muddy current, and at last, without his shield, partly swimming and partly wading, got across. Caesar and his company were amazed and came to meet the soldier with cries of joy; but he, in great dejection, and with a burst of tears, cast himself at Caesar's feet, begging pardon for the loss of his shield.
>
> (Plutarch 1919, para. 16)

This passage from Plutarch reveals many characteristics of heroism. Its declared motive is to illustrate the impact Caesar had on his troops, which was so substantial that, says Plutarch, it allowed him to win the great Roman Civil War. The passage involves a presupposition: a soldier is not worthy of any merit if he is found without his shield; to abandon a shield is to be exposed to an inability to self-protect, which leads to the inability to protect one's brothers-in-arms, and this in turn constitutes, ultimately, a failure in the soldier's ability to protect the emperor himself. The shield is the essence of the soldier's identity, as very well portrayed by Plutarch in his account of the brave soldier lying on his knees, seeking forgiveness for his failure in losing his shield, a mistake that negates completely the image of him as hero.

Similarly, today's soldiers strongly identify themselves with their ability to self-protect, as emphasized through their reliance on their military gear, weapons, and equipment. The ability to protect oneself, one's brothers and sisters in arms and ultimately to serve the mission, is embedded in the soldier's identity. These soldiers highlighted the significance of the Rules of Engagement

(ROE) in their right to self-protect as soldiers, compared with an absence of that right as police officers:

> Hum, I don't think I would, I'd be a policeman today. Not knowing what I know, not, not having done what I've done and not having trained the way I've trained, because, huh, conversations with my family members that are in the police service, they have to, they're, they are bound to put themselves at massively unnecessary risks, a lot of risks, hum, before they're allowed to do what they need to do. Hum, if you look at the media today, huh, certainly more popular stories about policemen, how they, they 'Tasered' this person or they, they pepper sprayed that person and, 'oh how could they do such a horrible thing?' Well, the policeman did it because his life was in danger and because he, he protected his own life; the media is going to tear him apart for it. It's really, really, I could not be a policeman any more. Not knowing what I know, not having done what I've done.
>
> (File 19)

> So, we're trained as – through battle school, I was training to, hum, basically to, you're trained to kill and you're, that's what your focus is. And then when we, hum, went to Croatia, you were getting shot at and you're seeing lots of horrible things but you can't do anything in return. So, it was, huh, it was pretty hard for a lot of the guys to kind of switch off like that. You had to really follow the Rules of Engagement and at that time, they were enforced very strictly.
>
> (File 61)

This survival instinct affects warriors to the point of psychological and identity collapse, when faced with events that make them unable to achieve what is perceived as the core of their profession. When a member is unable to fulfil this core role of the protector, it is as if the warrior has been stripped of his weapon and it is often self-interpreted as the soldier's failure to perform the mission because 'the weapon symbolizes protection of the identity of the soldier, and the means of defending all values'. (Scurfield and Platoni 2012: 306). For instance, in a context of voluntary self-disclosure about medical issues, out of the 75 interviews conducted with CAF soldiers who were deployed as members of the combat arms, 15 expressed some level of psychological distress (not deemed to be permanent):

> Because I came home I had to leave my friends at war, I didn't get to finish the tour myself which was a lifelong dream, um (...) I didn't, I didn't care any more. I was just, I got home and I, there was no option to go back and it looked like the tour was winding down. So, I had a lot of issues.
>
> (File 14)

Stripped of the confidence and pride soldiers exhibit after a victory or upon their return home, this soldier struggled to gain back his confidence in his role as a soldier (similar to the one described by Plutarch who begged for forgiveness). This identity crisis, although often temporary, is nevertheless quite frequent in the cases of soldiers who lose their physical capacity to do their job.[20] That this perceived self-failure in the role of the soldier and protector induces such adverse mental health outcomes illustrates the strength of an ideology: the importance of self-protecting.

Duty and chaos

To self-protect allows soldiers to focus on the mission as the first priority, with the welfare of troops second and with self-interest serving as the least important of the priorities.[21] If, in theory, duty involves displaying dedication, initiative and discipline while executing a task, in practice it tends to dehumanize the behaviour of the soldier who has to sacrifice family and personal needs for the success of the mission. For instance, basic training itself aims at dismantling one's personality so that it can be rebuilt into the ideal soldier identity:[22] a soldier who only focuses on the mission – as a team player. In training, when a soldier demonstrates a weakness or an injury, his teammates are required to carry on with their own tasks while also 'picking up the slack' of their weaker or injured teammate. Soldiers are taught from indoctrination that if an individual cannot self-protect, then he cannot protect his fellow troops or carry out the mission. This sense of commitment and cohesion is reinforced to the point where soldiers, who cannot protect themselves after an injury, feel they have failed their team-mates and consider themselves a failure in their core identity as a soldier:

> But the hardest part for me was actually getting sent home from it. That I had a really, a lot of issues with actually. Because I came home I had to leave my friends at war, I didn't get to finish the tour myself which was a lifelong dream.
>
> (File 15)

Similar to the ancient soldier having abandoned his shield, this soldier's injury rendered him incapable of protecting himself, and therefore unable to protect his brothers in arms. Without his metaphorical shield, which serves as the symbol and essence of the soldier's identity, his role as a soldier was challenged:

> Like when I am doing work up training to go overseas or I am overseas that's when I'm happiest, that's when I feel like I am doing my job. That when I'm, you know, fulfilled but uh, being in garrison and doing endless field exercises for no specific purpose is difficult for me.
>
> (File 15)

The comfort zone, for many soldiers, is war and its destructive effects on people and things. When soldiers' injuries force them to return home before they have completed what they were ordered to, issues start arising; they feel a sort of vertigo, as if someone had rapidly removed the floor on which they were standing from under their feet. This disequilibrium is not caused as much by the injuries as by the physical removal from their comfort zone.

Loyalty in a world of ambiguities

This sense that one has failed in one's duty as a result of their failure to self-protect, albeit through injury and due to no fault on the part of the soldier, is so strong that soldiers who are already struggling to recover from a physical injury also experience psychological consequences, as they attempt to reconstruct their own self confidence that has been lost: they feel they are not able to function in the way they should, as soldiers.[23] After this soldier was sent home due to a broken leg, he experienced bitterness and guilt for failing to complete the mission and leaving his crew behind:

> after I left, my crew sort of got switched around a little bit and the driver who came in who is a friend of mine ... he took over as driver and he ended up getting killed because they rolled over a, what they figure was an old soviet mine ... So that was actually after I left, but that kind of a, affected me because there is always the what if's ... . If I had been there, would I have the plough down,[24] would I have even been in the field, would any of that had happened, so... Mostly my issues with that tour wasn't anything that happened while I was there.
>
> (File 15)

The ethics required by a soldier in his commitment to his role (role morality) differs from those required of a civilian (ordinary morality).[25] The soldier's loyalty to his comrades goes beyond the influence of ordinary morality for a civilian; in a combat environment where a soldier's survival is rooted in his trust in his brothers in arms, morally culpable behaviour such as adultery, swearing, excessive drinking, etc., is subordinated to the more immediate factors related to their role as soldiers, 'or merely whether he is brave and knows how to use his weapon' (Robinson 2007: 25.) The loyalty that is required of soldiers obliges them to remain loyal to their country and thus to the entire chain of command. For this outcome to be fulfilled, such loyalty must be based upon mutual trust between the soldier following the orders and the commander issuing them. However, the policy makers and strategic commanders issue orders in accordance with the politics that motivate the Canadian government. These politics, rooted in the public opinion that favours Canadian peacekeepers, influence the orders being passed through the chain of command, eliminating terms that may cause controversy with Canadians, such as 'enemy':

> There were times where I was counselled in the Spring of 2006 not to use the word 'war', particularly talking to media, and to be careful not to use the word 'enemy' to describe the Taliban. I personally had a problem with that.
>
> (File 87)[26]

As a result, officers and senior NCOs in theatre were directed to lie to their troops in an attempt to portray the war they are fighting against an enemy they want eliminated as a combat mission of security and protection:

> Though some senior members chose to portray the war for what it was: I used the word 'enemy'. Because it was the only way I could operate. The only way I could function as a commander, trying to push a philosophy onto the soldiers, is to use the right words, the words which the soldier can identify with, because to the soldier that's the enemy he's trying to kill. And there isn't any gray in that, right? Therefore, they're in a war, because they're participant.
>
> (File 87)

The reality of the soldiers' war experience had them fighting against a different 'enemy' from that in the positive mission portrayed to the country by the higher chain of command and by the federal government:

> Since 2006 when the mission became combat-focused: in Kabul in 2004, we were not participating in the war. We were a stabilization force in the city. (...) It wasn't until later on that NATO expanded into a war role in the terrain of Afghanistan. So, it wasn't a war in 2004, in the opinion of NATO. But it certainly was in 2006 in Kandahar.
>
> (File 87)

This is not a mission focused on protection, education, and the involvement of Provincial Reconstruction Teams, but rather:

> In a war where you are actually fighting someone, you're a participant in the fight, not sitting between partners that were fighting. There is a designated enemy in Kandahar, and that enemy is trying to kill us. That makes all the difference in the world, because you are participating in the fight.
>
> (File 87)

The tension that these differences between the stated mission and the realities of military operations placed on the soldier's loyalty can be seen in their testimonials, as soldiers attempted to reconcile their allegiance to the country with their role as Canadian soldiers and with their lived operational experiences: 'Just wanting to, to do your job and wanting to make positive change in the country. To get rid of some more bad guys' (File 20). This shift in perception of mission objective from 'effecting positive change' to 'killing insurgents' is made many times throughout the testimonials:

> We were doing with the combat mission, an exceptionally amount of good. 'Hum, we are providing stability and safety for a large population' (...) 'Hum, and we killed a lot of bad guys [laughter], a lot.
>
> (File 20)

Regardless of this tension between the reality of the battlefield and the political ideologies, the soldiers express a strong sense of loyalty to their job as soldiers, and the associated allegiance to serving their country. When questioned on what type of civilian job they would be doing if they were not a soldier, they often described civilian jobs as being boring or mundane, compared to the excitement of their role as a soldier, even with the restrictions of the military's rigid chain of command: 'I'd be working in a building on a [not clear] or something. Like a steel [not clear] or milling machine or doing some other mind-numbing job' (File 20). In comparison, regardless of the dangers they face, the excitement expressed through their personal experiences on operations show their commitment and loyalty to their role as soldiers:

> So, that was pretty memorable, it was pretty close call (…) one of our civilian helicopters was coming in to drop off supplies and, huh, the, huh, the insurgents lit it up and we, huh, we were in a pretty good fire fight, that was pretty interesting.
>
> (File 20)

Although the ambiguity of the mission is apparent through their discourse, their loyalty to their fellow soldiers, Canadian and foreign alike, is seen across all the testimonials when describing their role, as if the dominant (indoctrinated) response is one of loyalty, regardless of the contradiction this strong stand point hides: 'Stop the insurgents from planting IEDs that are killing everybody, not just Canadians' (File 20).

Integrity in asymmetry

It is well known that in asymmetric wars, contradictions are the rule. To the contrary, during training, there are no grey areas: the mission and the soldier's role are crystal clear, black and white. Throughout the work-up training process in preparation for the deployments in Afghanistan, soldiers' mentalities were infused with the notion that they were deploying to help the citizens, bring peace, and kill the Taliban: 'And basically my job was to – well, protect the people and hunt down Taliban and kill them' (File 21). However, when sharing personal anecdotes, their discourse reveals an internal struggle of integrity from trying to account for their actions in line with these learned objectives. Senior members often expressed a clear vision of what the mission was meant to be – to bring peace and protect the citizens from insurgents – where it was deemed noble not to die in combat, but 'to help build a country, help the people'. However, the realities of theatre revealed duties and tasking that were ambiguous, with the consequences of the execution of these duties and tasking being for the most part intangible, unclear, or untraceable. Throughout his interview, a soldier referred to Afghanistan as a mission of protection and peace, yet his personal anecdotes were comprised of firefights with an invisible enemy:

'Cause I mean, if there was a real, if it was a real war, there'd be a general and he would have an objective. But there's no objectives 'cause there's no enemies 'cause they are ghosts.

(File 21)

For many soldiers, their deployment to Afghanistan lacked these clear object-ives. The chain of command, the ambiguity of the mission and of its identified objectives forced the leader of his troops to carry on firing with no notion of whom he is firing at or why: 'they were these sort of ghosts, these ghosts that we're trying to kill' (File 21). This officer was given the order to kill, but felt, along with his troops, that he was firing at ghosts in a mission that sounded like a bad fantasy.

A major challenge to the last-mentioned soldier's integrity was related to the accountability for his and his team's firing actions. In Afghanistan, the enemy is often embedded within the civilian population, and hidden behind or inside civil-ian infrastructure, making the targets difficult to identify:

They, they, you know, they go behind a wall and will shoot but sometimes they're, you know, they're a bit too close. You know they could be any-where from two hundred metres away to ten metres away. And if they drop their weapon, and just walked down the road, we can't identify them. They just look like a farmer, look like any other person.

(File 21)

This soldier compares the enemy to ghosts on six separate occasions, demon-strating an understanding of the abnormal and eerie characteristics of having an unknown and unseen target:

It's that the enemy has no identity. They're very difficult to find. Even when they're attacking us, we still couldn't see them. It was a very tricky affair because they were, as I say, they were like ghosts so, how do you hunt a ghost?

(File 21)

How can a leader model responsibility and accountability to their troops when both the mission and the target are ambiguous?

Even more so, in Afghanistan, the lack of clear military objectives sometimes resulted in an absence of formal orders:

I mean I guess in some ways it was a war but my picture of war is like sol-diers. You know, conducting tactics and operations and strategies. Afghani-stan we didn't have any, like I never got orders. I never got orders once. (…) Well what are they gonna order me to do? I was just in charge of a combat outpost and my job was to protect the people. I was, well I got orders once and that was it.[27]

(File 21)

Over the span of six months, this was his one and only series of orders, summoning him to protect the people. When a direction from a soldier's own chain of command is unrealistic for the situation, this creates a division between a leader's authority to command tasks for an ambiguous mission and his ability to inspire integrity and loyalty in his subordinates. The ambiguity of the mission for this soldier forced him to invent duties and taskings for his troops, 'It was difficult 'cause I had to pull everything out of my ass' (File 21). He would give orders to fire on an enemy when he wasn't certain where they were firing: 'You need to select a target even if you don't see the enemy, you have to at least shoot at something, as opposed to shooting indiscriminately without looking' (File 21). The ambiguity of their enemy caused his whole team to struggle with their own integrity.

> Like for some of the soldiers, they start to identify civilians as the enemy because they never see the Taliban. They just get shot at and they get blown up. They never ever see the Taliban. All they see is people and they know how they operate and they know that some of these guys are Taliban. So, they start to superimpose the, the face of the Taliban, you know, this, this ghost, this, this person with no identity onto civilians. (...) So, you start seeing everyone as the enemy.
>
> (File 21)

According to this soldier, they were challenged by the difficulty of distinguishing the enemy from the civilian and thus, had difficulties defining their own accountability, up to a point where they could not differentiate between whom to shoot and whom to protect. These disconnects between their roles as soldiers and the state mission caused considerably anxiety and discontent within the platoon.

Indoctrination makes integrity an important aspect of the soldier's ethos, causing tension and anxiety when it is not achieved. After discussing the difficulty he and his soldiers experienced with identifying the insurgents, he continually made references to the ROE, as if trying to accentuate the fact that the problems they faced were not of their own doing, as they were obeying the rules; rather, it was the ambiguity of the mission that created these tensions in responsibility and accountability. When they first came across a corpse, assumed to be as a result of their own fire, they felt relief to have evidence of the consequences of their actions: 'When we found that guy, we didn't even know we killed. That was one of the *best moments*' (File 21).[28]

This confirmation of their shooting blindly at the direction of fire answered two underlying questions; first, that they weren't firing at the Afghans they were meant to be protecting, and second that their fire was having an impact; they were actually doing their job of eliminating the danger:

> It was a great feeling. It was like scoring a goal in a big hockey game. It was immense satisfaction because you knew, there was no doubt, that this guy had just tried to kill you. And you killed him. And there he was, lying dead

in front of you. It was immensely satisfying for me and for the troops too. It was very good for morale. 'Cause that's the only time that you can really know that you've, you know, done something correctly.

(File 21)[29]

Because the soldier's training emphasizes the importance of the soldier's ethos, their identity as professional and ethical soldiers is threatened when they cannot prove to themselves that they have carried out their role in accordance with these values, regardless of whether or not this self-confirmation is realistic and attainable given the ambiguity of the stated mission.

Perversities of courage

The soldier's ethos, again according to the ethics manual, *Duty with Honour: The Profession of Arms in Canada,* has specific requirements for a degree of courage beyond the normalities of the civilian society. CAF members enrolled in the only profession that involved unlimited liability.[30] Throughout their careers, soldiers are trained to react to stressful situations quickly and effectively, temporarily relinquishing any normal, emotional responses, thoughts, or behaviours that might hinder this capacity to react in an ethical and professional manner. In a combat situation entailing life-threatening scenarios, fear becomes the enemy:

> there's a certain type of person that is just kind of wired for this. Hum, and I've always been wired for this. Hum, a dangerous, you see something dangerous, you hear something dangerous, a loud bang, there's a fire or whatever, most people move away from it. Hum, the odd few will move towards it to try and help or to try and, and, you know, do something about it.
>
> (P21)

> Soldiers are generally selfless people. Hum, they are, hum, they are willing to go and live in some of the most terrible conditions and be happy about it [laughter]. (…) you know, they're, they're an odd, they're an odd member of society. Hum, they're a type of person that, that moves towards danger as opposed to away from it.
>
> (P21)

> Hum, and I don't know, it's just the way my brain works. If, if there's something bad happening, if there's a fight, if there's, huh, hum, people in trouble, somebody getting hurt or fire, I'll move to it. It's just, it's instinct. It's, it's just the way my brain works.
>
> (P21)

Due to the unpredictable nature of actions in a war theatre, soldiers must follow their training and behave in an almost emotionless, robotic manner in order to carry out their mission and survive through the day. Without this level of courage, the mission cannot move forward; soldiers would be frozen by every physical and emotional tragedy. Finishing a task or completing a mission has to be the focus of the team. When this soldier witnessed an IED explosion and was informed that his close friend was hit, it took a great degree of physical and mental strength to pull himself and his men through the rest of the day's requirements:

> I went to the platoon commander and I said, 'Sir, don't ever, ever tell me who it was while we're on patrol again. Wait til we get back' 'Cause it was just too much of a detrimental. So, I mean, it would, it would practically take the wind out of your sails at the time of the incident.
>
> (File 19)

The reality of the dangerous environment is what helped him to suppress the normal 'civilian' responses to such a tragic event: mourning, questioning the situation, and responding emotionally. This soldier quickly realized the incapacitating effects of his normal emotional responses to trauma, and was able to move beyond them so he could focus on being a soldier.

He also shared a story that tellingly demonstrates what he considered to be the epitome of a soldier's capacity to respond and persist in the face of danger and fear.[31] In Afghanistan, when a soldier loses their life, a ramp ceremony is conducted as the bearers carry the bodies of the dead onto the plane to send them home, offering an environment of respect and mourning for the fallen soldier. After the first loss within this soldier's section, he and his crew missed the ramp ceremony to offer their own form of respect for their fallen comrade:

> I couldn't be there because we, my section went and did his patrol for him even though he was dead. So, we, we got mounted up in, in our vehicles, our little Jeeps and off we went. Huh, and we were actually quite close to, to the blast site where he was killed. And I remember driving around doing the patrol that, that sergeant should have been doing and the plane flew overhead and I remember we stopped, we took a moment. Hum, that was a very sad moment for all of us. But, although I couldn't be there to, to see my, my mentor off, hum, I, you know, taking up the fight, taking up his patrol and carrying on what he, what he would have been doing, was, huh, it was a badge of honour for me.
>
> (File 19)

Throughout this story, he expressed a strong underlying tone of admiration for his section for demonstrating the courage and strength of mind needed to take up their fallen brother's torch, and to overcome their feelings of fear and brokenness. The normal expectations of society would see these soldiers removing themselves from their duties in response to the trauma of such a loss, but the expectations of exemplary courage within the military requires those human responses to be temporarily suppressed for the purposes of operational effectiveness. For many soldiers, demonstrating the strength and resolve to carry on is a means of offering respect to their fallen brothers; their loyalty to their brothers in arms motivates them to persevere:

> because my friends have died, is no reason for me to stop. I have to keep going. Hum, they wouldn't stop if it was me that was killed, so I'm not going to stop because they're gone. And I'll pick up where they left off and I'll keep going. So, it was, it was a motivator I think, it was a motivator.
>
> (File 19)

It was not until after his return to Canada, and speaking with doctors, that this soldier understood that the courage he admired and strived to portray as a soldier was not considered normal for the civilian population:

> And, you know, the soldiers would come in and they'd sit in my office and I'd say, 'Ok well what's going on?' And they'd start telling me things and I'm thinking, 'This is me. I'm, I'm hearing my own story being told to me'. And at first I was like, 'well he's not sick, this is normal', but no, it's not normal.
>
> (File 19)

After talking to medical professionals about these soldiers' behaviours he came to the conclusion: 'Wow, I'm not well either. I'm not, I'm not acting in a normal way, I'm not acting the way most normal people do; I'm acting the way a soldier does but that's not necessarily normal' (File 19). This behaviour made him an exceptional soldier, but a broken civilian. His determination to persist beyond a 'normal' human's emotional capacity met the expectations of his training, but resulted in psychological damage since he was, in fact, a normal human merely supressing, ignoring, and neglecting these instinctive responses. Yet despite the realisation of the psychological impact of his time overseas, he persisted in his determination to challenge the fears associated with missions overseas, and to take upon himself the responsibility to do it:

> And the only answer that I can come up with, that, that makes sense to me, is I volunteer so nobody has to be told to. And if I don't volunteer first, who will? So, by, by volunteering, I make sure that, that I'm willing to go and do that. So, I'm making sure that somebody who's not willing to do that, somebody who doesn't want to go overseas, doesn't have to, isn't conscripted to go and, and do that, because, huh, somebody who doesn't want to be there, is not paying as much attention to the mission and to his duties as somebody who does want to be there.
>
> (File 19)

When questioned as to whether he would return to combat, even with the understanding of the psychological impact and the experience of the difficulties of reintegration, this soldier explained that he would definitely go back. Although one could argue for stubbornness or blindness to danger in this obstinacy, what is mainly revealed here is the need of soldiers to do their job. This soldier, like the 75 soldiers interviewed, could have argued in favour of going back to kill the Taliban, but they somehow all argued instead that they were 'fighting' to carry out their duty – their job – of helping others, revealing the strong influence of the soldier's ethos on the soldiers' thoughts and actions.

An organization such as the Armed Forces dictates behaviour in a way that will provide the necessary tools to its members to be able to defend themselves, to a certain degree, given the unpredictability of the actions of the enemy and the human factor, under the extreme circumstances of the war theatre. It also dictates that the same people restrain from any violent behaviour when not at war. The challenge for the CAF members while they are preparing for war or peace making, while they are in a mission, and when they are coming back to the regiment, lies between those two extremes: the very shield that first protected them turns against them, like the mythological Gorgon that was petrified by the very defence that it had used to turn its enemies into stone. The challenge for the CAF as an organization is even more extreme. The number of CAF members who consider themselves 'children of Afghanistan' – those who enrolled in the immediate prologue to and during this war – demonstrates the fundamental challenge to CAF training in terms of dealing with long-term identity crises and threats to well-being: how to maintain, when they are on garrison duty at home, that necessary shield of 'soldier identity' which kept them alive on the battlefield.

Notes

1 This chapter would not have been possible without the help of Mr. Jean-Simon Demers and Mrs. Melanie Belanger, my research assistants. This chapter is based on interviews sponsored by the Chief of Land Staff Research programme. The opinions expressed in this chapter reflect the opinion of the author and do not necessarily represent the opinion of the Canadian Army, the Canadian Armed Forces or the Department of National Defence.
2 *The Law of Armed Conflict at the Operational and Tactical Level*, Ottawa: Director of Law/Training, Office of the Judge Advocate General, National Defence Headquarters, last modified in 2012, electronic publication: www.fichl.org/fileadmin/_migrated/content_uploads/Canadian_LOAC_Manual_2001_English.pdf.
3 *The Law of Armed Conflict at the Operational and Tactical Level*, op. cit.
4 *The Law of Armed Conflict at the Operational and Tactical Level*, op. cit.
5 Chief of Land Staff, *Duty with Discernment: CLS Guidance on Ethics in Operations*. Ottawa: Directorate Army Public Affairs, 2009: 33, accessed from http://cradpdf.drdc-rddc.gc.ca/PDFS/unc104/p534069_A1b.pdf p. 22–23.
6 The Code of Conduct applies to all cf. personnel carrying out military operations other than Canadian domestic operations. These eleven rules are designed to allow our soldiers to successfully complete any military mission according to a standard of conduct demanded by our Army Ethos, our fellow Canadian citizens, and the international community in accordance with the Law of Armed Conflict:

   1 Engage only opposing forces and military objectives.
   2 In accomplishing your mission, use only the necessary force that causes the least amount of collateral civilian damage.

3 Do not alter your weapons or ammunition to increase suffering, or use unauthorized weapons or ammunition.
4 Treat all civilians humanely and respect civilian property.
5 Do not attack those who surrender. Disarm them and detain them.
6 Treat all detained persons humanely in accordance with the standard set by the Third Geneva Convention. Any form of abuse, including torture, is prohibited.
7 Collect all the wounded and sick and provide them with the treatment required by their condition, whether friend or foe.
8 Looting is prohibited.
9 Respect all cultural objects (museums, monuments, etc.) and places of worship.
10 Respect all persons and objects bearing the Red Cross/Red Crescent, and other recognized symbols of humanitarian agencies.
11 Report and take appropriate steps to stop breaches of the Law of Armed Conflict. Disobedience of the Law of Armed Conflict is a crime.

(*Duty with Discernment*, 2009: 33–34)

7 *Duty with Discernment*, op. cit., 13.
8 *Duty with Discernment*, op. cit., 23.
9 The 75 testimonies were collected by this chapter's co-author Stéphanie A.H. Bélanger, PhD, for the Research Report #2010–0333 – SLA: La culture de la guerre et l'identité du soldat/Warrior Culture and Soldier Identity, sponsored by J1 Pers, Chief of Land Staff. The testimonies quoted in this chapter are identified by 'File' followed by a number, in parenthesis, and are part of the 75 testimonies that were made by Bélanger, *Entrevues. Personnel* (collected October 2010–October 2011). Trans. Stéphanie Laszewski. The opinions expressed in this chapter reflect only the opinions of the author and do not necessarily represent the opinion of the Canadian Army, the Canadian Armed Forces or the Department of National Defence.
10 The relationship between ethics and identity, and the study of testimony as a means to analyse the impact of this relationship has been discussed in many scholarly and military publications such as: S.A.H. Bélanger, 'The Testimony of a War Amputee from Afghanistan: Discursive Myths and Realities'. in *Shaping the Future, Military and Veteran Health Research*, edited by A.B. Aiken and S.A.H. Bélanger, 255–268. Winnipeg: Canadian Academy Press, 2011; R. Amossy, *La Présentation de Soi: Ethos et Identité Verbale*. (Paris: Presses universitaires de France, 2010; C. Dornier and R. Dulong, *Esthétique du témoignage: Actes du colloque tenu à la Maison de recherche en sciences humaines de Caen du 18 au 21 mars 2004*. Paris: Éditions de la Maison des sciences de l'homme, 2005; J. Toiska-llio, *Identity, Ethics, and Soldiership*, Helsinki, National Defence College, 2004; P. Ramadanovic, *Forgetting Futures: On Memory, Trauma, and Identity*. Lanham, MD: Lexington Books, 2001; Stéphane Beaud, 'L'usage de l'entretien en sciences sociales. Plaidoyer pour l'entretien ethnographique'. *Politix* 9, no. 35 (1996), 226–257.
11 See Beaud Stéphane, 'L'usage de l'entretien en sciences sociales. Plaidoyer pour l'entretien ethnographique', *Politix* 9, no. 35 (1996), 226–257.
12 See Joshua Goldstein, *War and Gender: How Gender Shapes the War System and Vice Versa* (Cambridge: Cambridge University Press, 2001); Rudy Richardson, Desiree Verweij and Donna Winslow, 'Moral Fitness for Peace Operations', *Journal of Political and Military Sociology* 32 (2004), 99–113; etc.
13 See Kate McLoughlin, *Authoring War: The Literary Representation of War from Iliad to Iraq* (Cambridge: Cambridge University Press, 2012), 14; Philippe Lejeune, *Je est un Autre: l'Autobiographie de la Littérature aux Médias* (Paris: Seuil, 1980); Jean-Philippe Pierron. *Le Passage de Témoin: une Philosophie du Témoignage*. (Paris: Les éditions du Cerf, 2006); etc.
14 Howell and Yelle, *Your-Say: Spring 2009 Core Section Results*, 17.

15 This and other similar quotes references are taken from the author's own interview records.

16 *Duty with Honour: The Profession of Arms in Canada*. Winnipeg: CDA Press, 2005 (2003): 11, accessed from www.publications.gc.ca/collections/collection_2011/dn-nd/ D2-150-2003-1-eng.pdf which is the most recent joint publication from the Department of National Defence and Canadian Forces, *Code of Values and Ethics*, published by the Defence Ethics Programme, 2012, as per Annex A2/3 and Annex A3/3, with some minor distinctions with the Public sector (civilian) values (respect of democracy, respect of people, integrity, stewardship and excellence). When both military and civilian employees of the Department of National Defence are put together, the values become: integrity, loyalty, courage, stewardship and excellence (the difference being that duty, while enumerated in the values of CAF members, is removed here and replaced by stewardship and excellence).

17 Victor Davis Hanson, 1989, *The Western Way of War, Infantry Battle in Classical Greece*, London: Hodder and Stoughton: Chapter VI.

18 Homer, *Iliad*, any edition, Chapter 18.

19 A very popular example of this would be the end of the Second World War, a time of exhaustion and fear of losing against a fearful enemy: 'We're not going to just shoot the sons-of-bitches, we're going to rip out their living Goddamned guts and use them to grease the treads of our tanks. We're going to murder those lousy Hun cocksuckers by the bushel-fucking-basket' (General George Patton, 1944: East Anglia, quoted in Lapham's Quarterly, www.laphamsquarterly.org/states-war/ addressing-troops).

20 See, for instance, the feeling of failure expressed in A. Loyd, 2000, *My War Gone By, I Miss it so*, Toronto: Penguin p. 322 (on the War in Ex-Yugoslavia); see also L. Faulder 2007 *The Long Walk Home: Paul Franklin's Journey from Afghanistan*, Victoria, British Columbia: Brindle & Glass p. 184, which has been analysed in S.A.H. Bélanger, 'The Testimony of a War Amputee from Afghanistan: Discursive Myths and Realities' in *Shaping the Future: Military and Veteran Health Research* (Kingston, Ontario: Canadian Academy Press, November 2010, 265–268.

21 See R. Jepperson, 'Institutions, Institutional Effects, and Institutionalism' in W.W. Powell and P.J. Dimaggio, 1991, *The New Institutionalism in Organizational Analysis*, Chicago: University of Chicago Press, 147.

22 See P. Robinson (2007) and T. Nadelson *Trained to Kill: Soldiers at War*, Baltimore: Johns Hopkins University Press, 2005.

23 See Bélanger, 'The Testimony of a War Amputee from Afghanistan', 2011 (n. 20).

24 An instrument on the tank for removing mines, which was not engaged when his friend hit the mine.

25 See Robinson 2007.

26 Note: this interview was not conducted in the same context of the 75 interviews from which the statistical data are given earlier in this chapter.

27 Over the span of six months, this was his one and only order.

28 Emphasised by the authors.

29 Hockey was used by six of the interviewees in various metaphors to express emotions and experiences for which they lacked military comparisons and discourse, a comparison unique to the Canadian culture.

30 A-PA-005-000/AP-001 *Duty with Honour: The Profession of Arms in Canada*, 2009, Chapter 2, Section 2:

> Unlimited liability is a concept derived strictly from a professional understanding of the military function. As such, all members accept and understand that they are subject to being lawfully ordered into harm's way under conditions that could lead to the loss of their lives.

See also CFJP-01 – Canadian Military Doctrine, B-GJ-005–000/FP-001, Chapter 4:

Unlimited liability is the fundamental condition under which all members of the cf. serve. They are required to accept, without reservation, that they must carry out their missions and tasks regardless of personal discomfort, fear, or danger. Unlimited liability is the cornerstone of military service and distinguishes cf. members from their civilian counterparts.

31 According to Aristotle, 'It is when everything is lost that you can appreciate the courage of great heroes'. Aristotle *Rhetoric*, Chapter IX, any edition.

Bibliography

Amossy, R. (2010) *La Présentation de Soi: Ethos et Identité Verbale*, Paris: Presses universitaires de France.

Beaud, Stéphane. 'L'usage de l'entretien en sciences sociales. Plaidoyer pour l'entretien ethnographique'. *Politix* 9, no. 35 (1996), 226–257.

Bélanger, S.A.H. (2011) 'The Testimony of a War Amputee from Afghanistan: Discursive Myths and Realities' in A.B. Aiken and S.A.H. Bélanger (eds.), *Shaping the Future, Military and Veteran Health Research*, 255–268, Winnipeg: Canadian Academy Press.

Canadian Defence Academy (2003) *Duty with Honour: The Profession of Arms in Canada*. Winnipeg: CDA Press, 2005 online at www.publications.gc.ca/collections/collection_2011/dn-nd/D2-150-2003-1-eng.pdf.

Chief of Land Staff (2009) *Duty with Discernment: CLS Guidance on Ethics in Operations*. Ottawa: Directorate Army Public Affairs, online at http://cradpdf.drdc-rddc.gc.ca/PDFS/unc104/p534069_A1b.pdf.

Dawes, J. (2002) *The Language of War: Literature and Culture in the U.S. from the Civil War Through World War II*, Cambridge: Harvard University Press.

Dornier, C. and Dulong, R. (2005) *Esthétique du témoignage: Actes du colloque tenu à la Maison de recherche en sciences humaines de Caen du 18 au 21 mars 2004*, Paris: Éditions de la Maison des sciences de l'homme.

Faulder, F. (2007) *The Long Walk Home: Paul Franklin's Journey from Afghanistan*, Victoria, British Columbia: Brindle & Glass.

Goldstein, J. (2001) *War and Gender: How Gender Shapes the War System and Vice Versa*, Cambridge: Cambridge University Press.

Hanson, V.D. (1989) *The Western Way of War, Infantry Battle in Classical Greece*, London: Hodder and Stoughton.

Howell, G.T. and Yelle, M. *Votre Opinion /Your-Say: Spring 2009 Core Section Results*, (December 2011): 12.

Jepperson, R. (1991) 'Institutions, institutional effects, and institutionalism' in W.W. Powell and P.J. Dimaggio, *The New Institutionalism in Organizational Analysis*, Chicago: University of Chicago Press.

Loyd, A. (2000) *My War Gone By, I Miss it so*, Toronto: Penguin.

McLoughlin, K. (2012) *Authoring War: The Literary Representation of War from Iliad to Iraq*, Cambridge: Cambridge University Press.

Nadelson, T. (2005) *Trained to Kill: Soldiers at War*, Baltimore: Johns Hopkins University Press.

Philippe, L. (1980) *Je est un Autre: l'Autobiographie de la Littérature aux Médias*, Paris: Seuil.

Pierron, J.-P. (2006) *Le Passage de Témoin: une Philosophie du Témoignage*, Paris: Les éditions du Cerf.

Plutarch (1919) *Life of Caesar*, Vol. VII of the Loeb Classical Library edition.

Ramadanovic, P. (2001) *Forgetting Futures: On Memory, Trauma, and Identity*, Lanham, MD: Lexington Books.

Richardson, R., Verweij, D. and Winslow, D. 'Moral Fitness for Peace Operations', *Journal of Political and Military Sociology* 32 (2004), 99–113.

Robinson, P. (2007) 'Ethics Training and Development in the Military', *Parameters*, 37(1), (Spring 2007), 23–47.

Scurfield, R.M. and Platoni, K.T. (eds.). (2012) *War Trauma and Its Wake: Expanding the Circle of Healing*, New York: Routledge.

The Law of Armed Conflict at the Operational and Tactical Level (2012) Ottawa: Director of Law/Training, Office of the Judge Advocate General. National Defence Headquarters, online at www.fichl.org/fileadmin/_migrated/content_uploads/Canadian_LOAC_Manual _2001_English.pdf.

Toiskallio, J. (2004) *Identity, Ethics, and Soldiership*, Helsinki, National Defence College.

Winslow, D. (1998) 'Misplaced Loyalties: The Role of the Military Culture in the Breakdown of Discipline in Peace Operations', *The Canadian Review of Sociology and Anthropology*, 35(3), 345–367.

6 An organic professional military ethic and the educational challenge

Sally Rohan

Educating the military in the ethics of their profession, clearly demonstrated by the various contributions to this volume, is at best a contentious pursuit and one that falters frequently in the area of definitional clarity in, and consensus on, the prevailing lexicon. One area of definitional vagueness is the notion of a Professional Military Ethic (PME), which might, in its broadest sense, be seen as encompassing the relationship between the profession of arms and the principles and values that provide its moral basis. Fundamental to any definition of this Ethic is an understanding of the nature of the Armed Forces as a profession, and of how moral principles are then derived and inform that professional existence. In this analysis, the relationship between the military and their Ethic is explored in an effort to provide a better understanding of the part that this construct may play in the ethical education and development of the professional military. In so doing, it seeks to encourage a practical understanding of the Ethic as a tool of professional socialisation, resting on moral principles, but also grounded heavily in the functionality of the service.

The central contention of this chapter is that the PME may provide both a moral and ethical foundation for behaviours that are intrinsically tied to the concept of professionalism. For the purposes of this study, the PME is articulated in terms of four broad conceptual elements. The first element, the moral authority of the profession, derives from the concept of *service*. The second and related element comes from the military's *moral purpose* – what it is for (why it fights). The third element comes from *moral expertise* – how it fights – and is influenced by domestic and international law along with societal norms and expectations. And the final element relates to *moral community* – the basis of the corporate identity and culture of the 'learning organisation' that the professional serviceman inhabits. Collectively, identification with, and orientation towards, each of these four fundamentals, as interpreted and understood in context, may provide a universally applicable foundation for the practice of ethical education for national militaries.[1] Most significantly, where moral ambiguity and appeal to duty or character may not be sufficient, persuasive or applicable, the idea of what it is to be a professional – a member of a functional as well as ethical community, bounded within this concept of the PME – may provide a touchstone for right actions. Grounded in the appeal of 'belonging', this Ethic becomes both immediate and accessible, and potentially more conducive to ethical behaviour formation than esoteric appeals to personal or universal standards of military 'goodness'.

Interest in the concept of a PME, along with broader efforts to enhance professional ethics education, reflects widely held concerns regarding the slippage of socially expected ethical standards within the military community. In this context the establishment, codification or inculcation of the Ethic has been identified as a route to rehabilitation. In 1978, US Army General (retired) Maxwell D. Taylor's ponderings on the requirement for a formalised Military Ethic with its 'writ running from top to bottom of the officer corps', was a reaction to popular 'moral uneasiness' over a now strikingly familiar 'national fall from grace', apparent for the military by 'officer dereliction in Vietnam, and scandals at West Point' (Taylor 1978: 18–21). Thirty years later that call was echoed by General George W. Casey (Casey 2009), Chief of Staff of the US Army who, fearful of the perceived moral decline of his service, found substance in the tasking of a new centre at West Point to explore and promote the moral and ethical foundations of the army profession. Over the last decade, social disillusionment with what has been widely perceived as declining moral and ethical standards evidenced in areas of both the private and public sector, including banking, politics, the police and the media, has led to increasing cynicism regarding the professionalism of many of society's primary service institutions. After many years of active engagements in Iraq and Afghanistan, the Armed Services have not escaped criticism: whilst public support for those military personal serving on the front line has remained high (Hines *et al.* 2014),[2] there has been growing concern both within and outside of the military over perceived ethical failings at all levels of the military establishment. As the events in Abu Ghraib became graphically engraved on the public consciousness, the Gage and Aitken reports of 2011 into abuses by British military personnel serving in Iraq only served to confirm media stories of appalling transgressions of the socially acceptable norms of military conduct. Confined to a small minority of serving soldiers, these incidents were, sadly, also indicative of a wider problem of moral transgression witnessed across sister military institutions around the world, in the abuse of prisoners and civilians on operations and in detention, in the bullying and sexual harassment of serving personnel by their compatriots, and in the inglorious hedonism of senior officers who fell far short of the traditionally applied descriptor 'gentlemen'.[3] For many both within and outside of the military institutions, the all too frequent failure of the Armed Services to meet the exacting ethical standards expected of them pointed to a serious rot eating away at the foundations of the profession. These perceived ethical failings have led to calls across the democratic world for the promotion of military values and standards, and for the ethical education of serving military personnel.[4] The focus on the development of a PME as a means to develop 'right serving' military professionals has found most traction in the US, but it remains at best poorly understood and even more poorly articulated throughout the wider profession.

Defining the PME

The field of military ethics draws largely on pre-existing concepts of objective and foundational human morality as it applies to the military's role and behaviour in its use of force.[5] This moral universalism approach, although contested, draws on basic assumptions about the military condition and its relationship with widely held moral principles. If only all militaries could truly understand the moral basis of their shared professional existence, argue the moral purists, all would be well. However, this raises some challenging issues, not least for those who seek to translate ethical theory into military practice through the education of the professional membership. The first, and perhaps most daunting of these is that, whilst the recognition of a common foundational basis of ethical principles underlying the profession of arms may be widely accepted, the articulation of these principles in terms of a universal code of behaviours that has application across the military profession is less evident. Indeed, not only is there an absence of a single PME to guide right behaviours for the global community of military practitioners, there also appears to be little common agreement on the very nature of such an Ethic and what it should entail.

In seeking to clarify the nature of a PME, let us first rehearse briefly what constitutes a profession, for without the profession there can be no related professional ethic. A profession provides, develops and maintains a desired or necessary service to society that requires specialised knowledge and skills. When acting in their professional capacity, practitioners are defined by their 'role-differentiated behaviour', by the application of skills and the performance of activities to achieve the prescribed social 'good' (Hartle 2004: 6–9).[6] Effective professional practice requires individual autonomy of action and the application of expert judgement. This knowledge is based on both reflective and forward-leaning theoretical and practical knowledge, promoted by the attributes of an innovative 'learning organisation' (Senge 2006) responsible for regulating the behaviour of its members.

As unique subsets of society, professions establish their own subcultures, based on high standards and ethical codes, designed to promote socially approved effectiveness. In professing oneself to a particular role, the individuals within the profession are granted special rights and duties to apply their skills in relation to the social good that their expertise satisfies, just as they are also constrained in the use of these skills in order to retain public trust and hence their freedom of action. So, just as a surgeon may use a scalpel to perform a lifesaving operation, she may not wield it to do intentional harm. Unless behaving in a manner consistent with these professional expectations, an individual is not 'professional', even if they have expertise in their area of specialism. These rights and duties are supported by values and principles – what Bell (2012: 11) refers to as 'excellences' – which themselves derive from the requirements of the socially defined good served by the profession, and which further determine the acceptable ways and ends of practice and differentiate that practice from the established conditionality of normal social life. In this sense, they are the basis of what may be described as a professional 'role morality'. Often codified in core value sets and codes of conduct, and frequently finding expression in law, these articulations of conditional attributes and behaviours constitute the ethical parameters of the profession, not only guiding the practitioner in terms of responsibilities and duties, but offering what Carrick (2008: 197) refers to as 'moral self-defence' when required to perform acts outside of the established social norms of moral behaviour. Self-regulation, inherent to the notion of professions, ensures member compliance with accepted behaviours, be they formalised in written codes or permeated through informal standards and expectations, and is the task of the professional body and a condition of its autonomy (Hartle 2004: 29). Therefore, professionalism is an idea related to expertise, to skills and to the satisfaction of functional requirements – but it is also something more, including an expectation of self-governed standards and behaviours, which emerge from an understanding of the perceived professional good to be achieved by the service provided.

And so, this brings us to the PME. In practical terms, the notion of a PME is often used rather vaguely to encompass everything from military ethics (in all of its broad 'universalism'), to ethos (the 'spirit' of an organisation), to international Laws of Armed Conflict (LOAC), and to codes of conduct and/or values and standards (national and/or service specific) as they apply to military life. Although finding expression in these various forms (US Army 2013: 2–3) there appears to be no doctrine or widely recognised definition that clearly articulates the idea of a PME in a generally applicable manner. Even in the US, where the concept has received most attention, Moten (2010: 20) has described the Army Ethic as 'akin to the British constitution – it exists in a variety of forms, but it is hard to get one's arms around it'.

Given the laxity with which the term 'Professional Military Ethic' is employed, it might be useful to begin by taking it apart. The Oxford English Dictionary (OED) defines 'ethics' as 'moral principles that govern a person's behaviour or the conducting of an activity'. Hence, 'professional ethics' might be best understood as those moral principles that govern the behaviours of a person when carrying out their professional activities, and 'military ethics' as the moral principles governing military behaviours. Adding further clarity, the OED defines 'an ethic' (noun) as 'a set of moral principles, especially ones relating to or affirming a specified group, field, or form of conduct'. This suggests that the PME might be defined as something both broader and more specific than a set of foundational moral mores relating to military activity. It also points to the significance of group and conduct 'affirmation': it is this that opens the door to a more nuanced understanding of the Ethic as a relationship of parts, moral and functional, specifically applied to context. As such, grounded in the wider moral landscape within which the profession resides, the ethic is a representation of a unique moral community, central to this community being the spirit of the service and the identity to which its members adhere. When Snider, Oh and Toner describe the US Army's professional Ethic as 'a shared system of beliefs and norms, both legal (codified) and moral, which define the Army's commitment to serve the nation' (2009: 10), they begin to capture this broad reach of the Ethic as being beyond a set of moral principles and embedded in the concept of community within which serving soldiers, sailors and airmen identify themselves.

The debate regarding the purpose of the PME has revolved around the largely unhelpful distinction between its functional and moral aspects. For some, the emphasis is on what might be described as the 'fundamental moral guidance' role of the Ethic, as an expression of pre-existent moral ideas moulded to the specific requirements of the profession (Imiola and Cazier 2010: 11). Efforts to ground the PME in an objective morality are worthy, but the significance of the PME is that it has specific application to an area of professional competence that is, by definition, at least partly differentiated from the normal mores and standards of public life. The professional soldier is required to conduct him/herself according to ethical expectations that do not necessarily conform to standard practice outside of the profession. As Toner (1995: 18–19) notes, the PME is not a universal ethic, not even for officers in activities unrelated to their profession: it places emphasis on 'certain virtues, not for their intrinsic value for all men, but for their contribution to the formation of officers capable of performing their duties successfully in an environment of conflict'. Toner's code adopts a highly functional approach, 'he is a good officer to the extent that he succeeds, a bad one to the extent that he fails. His actions are right insofar as they contribute to mission success, wrong if they contribute to failure' (ibid: 20). The 2005 US Field Manual (FM) 1 *The Army* states that 'Professions create their own standards of performance and codes of ethics to maintain their effectiveness'. If the Ethic is based around the creation of standards and codes of professional behaviours that are driven by specific functional requirements, (in themselves removed from the realm of moral objectivity), one might ask whether this concept of an Ethic represents anything 'ethical' at all, if its purpose is simply one of efficacy. But it is evident that no military service identifies only 'expediency' as its aim. The functional benefits of complying with ethical standards may be many,[7] but being a member of a professional military does not equate to simply being an effective killer or automaton – it also has an ethical dimension. The military is a profession not solely because it is expert at its function, but because it upholds society's values, which it must exemplify and more – its 'excellences' that give meaning to the profession. As such, the moral component of fighting power is not just an expression of the contribution of moral factors to efficiency, but also of the significance of the military as a moral community, where purpose and method has a moral context and cannot be removed from it. Seeking to establish some sanctimonious moral wedge between service effectiveness and right intent seems to deny the significance of the PME as an expression of both a moral and a functional community.

In fact, a PME might best be understood as a vehicle for achieving a socially defined good, which has a functional and ethical context related to professional expertise and purpose, and which in turn confers specific legal and moral 'rights'. The purpose of a PME is to provide clear guidance as to what is right and proper behaviour for serving personnel, grounded in a concept of what it is to be a member of the profession. In establishing the moral principles that underlie the profession, it provides more than a broad-based and abstract moral code of right conduct, or a simply functional list of professional requirements. Rather it should seek to tie the moral foundation of the profession to the practices and purposes that the profession is founded to achieve. In this sense, the PME provides the source of moral identity at the heart of the profession – it is the ethical touch plate, expressing in ethical terms what the profession/professionals should be (the values that they embody), what they should know (the values that they defend and promote) and what they should do (to defend those values – how they should fight in a principled way).

PMEs are neither universal in application, nor static in nature, although the fundamental moral principles on which they draw may be enduring. The Ethic of any military institution, as a socially specific entity, must include reference not just to what is required by its underlying purpose (which itself may change over time and place), but will incorporate a way of being, thinking and doing that reflects the moral fundamentals of the society and state that it serves. That is not to support the idea of moral relativism, but to identify that the specific Ethic of any professional body, (and its constituent institutional representations), will be influenced by a range of factors that are universal, intra-professional and localised. Bounded by pre-existent moral beliefs regarding the right nature of human behaviours, the Ethic needs not only to be articulated in a way accessible to the 'unique audience' of the profession (Imiola and Cazier 2010: 11), but to reflect the particularities of the national military and service context, not necessarily readily transferable to other parts of the professional body.

A PME, in order to retain relevance as a set of guiding principles, also needs to reflect changes in the physical and normative environment in which the profession operates. Processes of military transformation and repositioning, as seen in the US Army post-Cold War, post-9/11, and now accompanying the transition to post-Afghanistan contingency, may result in dislocation and disorientation and a profound sense of loss of self-knowledge (Cook 2004: 56–58). The PME, as a representation of the body and spirit of the profession, must adapt to these changes in understanding of where and how the military acts, and for what purposes, if it is to give ethical guidance to an evolving profession. And this will be a challenge. Moreover, if a shift in professional jurisdiction (in terms of the skills and knowledge required in order to function effectively in new kinds of operations) has challenged the Ethic of a transforming US Army, then surely it must follow that national, or even service-specific professional military communities must be sustained by a particularistic professional Ethic, reflective of their understanding of service, purpose, mission and identity. For these reasons, the PME must be an organic thing, its relevance as a handrail for ethical being, thinking and doing being dependent on its ability to reflect the physical and normative environment that the profession inhabits.

A conceptual basis for PME

The PME provides the explicit and implicit guiding principles of right and proper behaviour for members of the profession in the conduct of their expert role. These principles derive from a set of concepts that define the profession both as a functional and a moral community, where the interplay of effectiveness and ethicality is fundamental to the idea of the profession. These defining concepts may be identified as moral authority, moral purpose, moral expertise and moral community. Variously understood and expressed, these concepts exist for all militaries, are intrinsically inter-related, and collectively constitute the domain of the PME.

The concept of moral authority

At the heart of the PME is the concept of moral authority (articulated in the idea of service) and its derivation. From whom or what does the military derive its moral right to be? The simple answer might be that it is the state that legitimises the military's use of force and, through the authority of the state, military personnel may (and at times must) inflict physical violence against people and property in a manner that would be prohibited within the norms of civil society. The military's special position as an agent of the state derives from the sovereign state's monopoly of the legitimate use of violence, formalised in the Charter of the United Nations and inferred in the traditions of Just War. This monopoly is, in turn, derived from the state's role as the agent of the people (the nation or citizens that it serves), the rights and interests of the individual citizen being represented and preserved through the political collectivity of the state. Hence, the authority of the military is based in the moral foundation of human rights, expressed through the political collective of the nation and represented by the state. This concept of the source of the military's moral authority is clearly articulated, for example, in the 2010 US Army White Paper, *Profession of Arms*, which identifies the moral basis of the Army Ethic as deriving from its defence of the US as a 'political nation that protects and respects human rights'. It is this moral authority that distinguishes the military from the 'factual' claims of such groups as drugs cartels and Private Military Companies, who might equally claim the use of violence to protect a way of life, or to act in common defence, but lack the political and moral authority of the state – the universally recognised (if sometimes contested) political expression of collective individual rights.

This raises the question: is the sovereignty conferred by de jure statehood a sufficient basis for legitimising military service? Where the state fails to represent and protect the interest of the people, the state itself may lack the moral legitimacy to confer authority on its military agent. This issue of the true source of legitimate authority of the Armed Services is tested in autocratic systems, where armed forces may be widely perceived as agents of citizen repression, although, even in the most oppressive of states, common appeal is made to the notion of a 'people's army', coming from the people and acting in the interests of the nation that it 'serves'.[8] If it is ultimately the people that confer moral legitimacy on the military, is there ethical legitimacy in military coups to displace autocratic (but sometimes democratically elected) rulers in the interests of defending the rights of those citizens? In identifying the root source of military authority, might the Westphalian relationship of conferred legitimate authority from state to the military be temporarily displaced by the over-riding needs of the citizens? This is an existential challenge for the military, whose legal existence, if not its moral one, is traditionally defined in terms of state agency: indeed, it is this that distinguishes the military from other forms of violent actors.

The subordinate relationship of the military to the state and to the society that it serves, and its subsequent obedience to civil authority, has provided the benchmark of military professionalism. The specific and desirable nature of state control has been contested, most notably in the classic works of Huntington and Janovitz. Assumptions about the 'right' nature of civil–military relations will be variously defined dependent on political, social and historical circumstances (Burk 2002; 2005). Acknowledgements of military subordination to the civil power may be derived from formal constitution or statute, from military oaths of allegiance, and from the loyalties articulated in the values and standards of the services. These statements of military allegiance, which in effect define the military's conception of conferred legitimacy, tend to specify, first and foremost, the nation, or its representation in the head of state and/or the state constitution. This is in contrast to other professions, such as medicine for example, where the professional's relationship to the individual (in this case, the doctor–patient relationship) excludes wider reference to government or office (Olsthorn 2011: 85–86). Where the political authority of the state, or its governing representatives, is weak or challenged, the military may appeal to more permanent or pervasive forms of 'authority' and legitimation. For example, in August 2013, in a symbolic expression of the military concept of service in a state recovering from its most recent post-democracy coup, the Egyptian armed forces revised their military oath on enlistment, omitting allegiance to the President in favour of 'the country, obeying the law and the constitution' (Gulhane and Taha 2013).

Loyalty to the military service is frequently directly identified as an extension of these primary commitments: the British Army, for example, whilst swearing allegiance to the Crown, acknowledges that the 'Nation, the Army and the chain of command rely on the continuing allegiance, commitment and support of all who serve' (British Army 2015). Royal Navy personnel, who unlike their peers in the other UK Armed Services, do not swear an oath of allegiance to the Monarch on enlisting, are reminded that 'The Nation, the Naval Service and those with whom we serve' are their focus of loyalty (St George 2011: 19), whilst US soldiers are urged to 'Bear true faith and allegiance to the U.S. Constitution, the Army, your unit and other Soldiers' (Seven Core Values). This concept of loyalty, whilst identified as a cardinal virtue in most militaries, gives rise to challenges within the organisation, as service, unit or comrades may more readily attract personal loyalty over and above the higher calling of professional loyalty to the source of the profession's legitimacy, or to the moral principle that underlies it. As Brigadier Robert Aitken observed, misplaced loyalty (typified by the 'wall of silence' met during the Baha Mousa inquiry) is a military tendency 'exacerbated in an organisation that trains its people in the virtues of loyalty, and which stresses the importance of cohesion' (British Army 2008: 24). Loyalty to the group over the source of moral authority must be seen as a professional failure. A clear articulation of the source of that authority and the implications for military service is a prerequisite for an effective PME that confirms the legitimacy of the profession through reference to its primary authorising agent.

Of course, authority in itself is neither moral nor otherwise: obedience to an authority that is corrupt or misjudged in its application of power is not a strong foundation for a PME that draws together the moral and functional in its commitment to a 'right authority'. Identification of that right authority and those principles that underpin it, however, enables a level of ethical clarity that should not be uncritical, but which offers a first point of departure for a consideration of right behaviours in support of right objectives.

The concept of moral purpose

The relationship between moral authority (who do we serve) and moral purpose (to what end) – the second of the conceptual areas of the PME – is clearly a symbiotic one. The specific purpose of the military may be defined in terms of its key role and missions – the objectives of its professional actions expressed through its engagements. What is the desired moral good that provides the military with its raison d'être? What is the military there to do? At its most fundamental, these questions might be answered by reference to the first concept – it is there to serve. But the 'how' of its service goes beyond that foundational assumption, and finds its ethical expression in what has been defined elsewhere as the 'moral ideal', the central and defining provision of the profession, which is the defence of the nation and its legitimate interests by threat or use of force (Davis 2003: 449).

The military are required to be prepared to give their lives for a higher purpose and that purpose has a moral and functional existence. The UN Charter, the LOAC and the Just War tradition provide the moral and legal basis of just military action. However, the challenge may be to determine not only what is permitted, but what is considered right purpose for any given military in relation to the values and interests of the people that it serves. The military's legitimacy requires society's acceptance of the profession's role and constitutional sphere or jurisdiction, but this is not assured and may be challenged by a range of situational factors. These factors will include the nature of the threat environment (the perceived risks), norm shifts regarding rights and interests (for example, from non-interference to Responsibility to Protect), and the introduction of new technologies (such as unmanned robotic weapon systems including armed drones) and methods (including asymmetric and terrorist tactics). At the same time, the civilianisation of areas of the military's role challenges traditional notions of professional jurisdiction, as skill-sets, tasks and capabilities are delegated from military to civilian operators.

The term 'mission' is commonly used as an expression of purpose that justifies a military's existence and privileges, although Wilson (2008) notes that the formally outlined tasks of Armed Forces 'may not capture the more profound sense of purpose that emerges from historical experience, where a history of political intervention on behalf of elite, military or citizen interest may influence current perceptions of legitimate action and behaviours'. Although the primary role of all militaries remains territorial defence, other roles may emerge, including anything from internal repression and coercion on the one hand to liberal interventionism in the pursuit of universal rights and freedoms on the other.[9] Whilst all militaries (and their political 'masters') seek to justify their role in terms of a 'moral' purpose, this may be challenged in interpretation or intent, and will require justification to those deployed, to the citizens of the state that deploys them, to allies and partners, and to the broader international community. Indeed, in counter-insurgency and stabilisation operations, perceptions of military legitimacy by those in the area of deployment, and by the wider networked audience of spectators (who might be influenced by the narrative of the engagement), may be fundamental to operational and strategic success (US Army 2014: 1–8; Chadha 2013). If the primary purpose of military service is to protect the rights and interests of national citizens or, as Taylor suggests, to provide the 'ultimate safeguard of the Constitution, our national institutions and the rights and benefits which we [sic] Americans are privileged to enjoy', (1978: 19) it might well be problematic to present such national considerations of purpose in moral terms. Where citizens' rights are themselves based on foundational moral values of human rights, it would seem to follow that a moral purpose for military action must take into account the impact of that action, (particularly where the objective may be perceived as the pursuit of national privileges), on the fundamental rights of others. Moral purpose requires just cause, and the articulation of that cause in terms of national interest has declining moral legitimacy for a global audience.

In his speech to the Royal United Services Institute in London in December 2013, the UK's Chief of Defence Staff noted that, whilst the Armed Services had never, in his 40 years of service, been regarded so highly, 'the purposes to which they have most recently been put has seldom been more deeply questioned' (Houghton 2013). This acceptance of purpose has a moral as much as a functional basis. Where the military fail to meet social expectations of success, where they are deployed to purposes that have limited popular support, when their politically sanctioned or directed methods (e.g. Guantanamo, waterboarding, extraordinary rendition), or military standards (e.g. Abu Ghraib, Marine 'A') fail to meet the values of the society they represent, the role of the military is inevitably called into question. The identification of purpose and the moral principles that support it must be at the heart of any conception of a PME, for it is purpose that gives meaning to the profession and the moral foundation of that purpose provides the moral compass for the pursuit of military action in complex, hybrid and often changing conflict environments.

The concept of moral expertise

Military personnel are professional where they demonstrate competency and expertise in carrying out their professional duties to provide a service to society that society cannot otherwise provide for itself. The concept of duty, derived from moral service and purpose, presupposes knowledge and understanding of the expert requirements of that service: that which needs to be known and those skills to be developed. Case *et al.* (2010: 3–4) identify four major areas of professional military expertise in their discussion of the US Army – military-technical, political-cultural, human development and moral-ethical. The Ethic inhabits this latter realm, which directs the military in its deployment of the rest of its expert knowledge in order for it to 'fulfill the fundamental duty of the profession to fight wars and conduct operations morally', in accordance with the expectations of the people and the requirements of domestic and international law.

What separates the military profession from all others has been bluntly expressed by Toner (1995: 22–23): 'in addition to killing and being prepared to kill, the soldier has two other principal duties ... some soldiers die and, when they are not dying, they must be prepared to die'.[10] In being prepared to inflict harm and take life, and to accept the 'unlimited liability clause' of giving their lives for their country, military professionals cross the most fundamental of moral premises, the right to life. But they must do this in accordance with the values and interests of the society that they support, and those moral principles that underlie them. The 'license' to cross these moral boundaries is dependent on clearly defined constraints on what are deemed morally permissible behaviours in relation to necessary moral ends, which establishes high expectations of moral knowledge and skills in the prosecution of professional activity.

It is in the realm of moral expertise that the question of professional autonomy is most clearly raised. The military, as a profession, defined in part by its ability to make judgement within its area of expert knowledge and skills, has a duty to provide those it serves with expert advice on the use and threat of force. In moral terms this involves resisting obedience to those civil proposals and policies that promote the unprincipled use of force (what Cook (2004: 63) describes as 'a high manifestation' of military professionalism) and exposing them to policymakers through expert and plain-speaking guidance (Cook 2013, 33–44). However, public confidence, not least in the UK and US, in the appropriate utilisation of expertise and knowledge by military professionals has been challenged by reports of inappropriate government lobbying by senior and retired officers, and questionable military relationships with the media and defence industry (Haynes and Pitel 2012: 9), facilitated in the latter case by the 'revolving door' of appointments for retiring military officers in the private sector. Clarity on the acceptable utilisation of professional knowledge is part of society's bargain with the military and is underpinned, for liberal democracies at least, by the public's belief in the 'apolitical and non-partisan ethic of service' and 'the principle of civilian control' (Moten 2010: 17).

Where states are party to international treaties and conventions regarding armed conflict, serving military personnel are duty bound to comply with derived legal obligations, the violation of which may result in prosecution through national or international courts (Waters 2011). A clear understanding of the legal constraints and permissions on military actions is thus an expert requirement of the profession. Where service personnel fail to act in accordance with what is held legally acceptable, they demonstrate a lack of expertise for which they may be held professionally and individually accountable. Military and Service Law and Administrative Actions provide further rule-based constraints on military behaviours, whilst Rules of Engagement impose tactical and operational limits on national forces. Collectively, these may be expressed as sets of basic rules for soldiers to follow.[11] Interpretation of legal or rule-based requirements into ethical behaviours, however, requires an understanding that goes beyond knowledge of what is officially permitted; it necessitates an awareness of ethical principles and an ability to apply them to decision-making (Hartle 2004: 102). In rapidly changing and complex conflict environments, the laws of war may be insufficient to indicate the proper behaviour of the soldier in a particular circumstance. National policy decisions on combatant classification, suspect rendition and interrogation have all challenged traditional clarity on the norms and laws of war and the subsequent acceptable behaviours of military personnel (Economist 2013). Nevertheless, a familiarity with the law's underpinning moral principles may provide that guidance to correct action beyond the *can or cannot* of rules to the *should or should not* of ethics (Wolfendale 2009: 61).

An understanding of the legal and ethical restrictions on action in relation to the military goals of any given operation will assist the military professional in making principled choices. Ethical decision-making requires a sound understanding of the moral goal of an operation, what values it is intended to secure, be it in defence of the realm or responding to humanitarian crisis. (Lucas 2003) This moral goal should be defined in operational intent (what is to be achieved), and reflected in operational planning for the ways and means of its achievement. Serving personnel will need to be cognizant of the acceptable use of force in relation to moral cost – what values are being protected, which can be subordinated and to what degree – and must, therefore, be educated in the translation of moral purpose into actions. And this is all the more the case in increasingly complex or hybrid conflict environments, and where the values defended (e.g. human rights) may appear to run counter to the violent acts that are required by the military objective (e.g. security). Modern conflicts pose extreme challenges for compliance with ethical and legal norms even where they are understood. Developing moral expertise in serving personnel provides the ethical 'shield' against immoral acts when fighting an enemy that neither recognises nor complies with the rules of war, and even in the absence of likely punishment for non-compliance.[12]

The most recent US COIN doctrine clearly articulates the relationship between ethical action and professional expertise in both moral and functional terms: 'Army and Marine leaders should clearly understand how adherence to a professional ethic provides the moral basis for unified action and how it becomes a force multiplier in all operations'. (US Army 2014: 1–10). Ethical competence, like other areas of professional expertise, must be developed in members of the military profession: just as the knowledge and skills required for the use of firearms or for the conduct of campaign planning requires education and training, so too must serving military personnel be prepared for the ethical challenges of their profession. The military profession should provide the learning organisations, facilitating and promoting the enhancement and exchange of knowledge and skills though its education and training centres, and through the interaction of practitioners and academics in the development and publication of doctrine and concepts relevant to the profession. The articulation of the realm of professional moral/ethical knowledge, through codes of conduct and 'values and standards', is unlikely to be sufficient in itself to prepare members of the profession for the moral challenges that their engagements may entail. As an articulation of the required moral expertise of the professional military body, the PME should thus offer guidance on the key moral skills and their application in the pursuit of the legitimate aims of military action.

The concept of moral community

It has been argued that, in the absence of a universal moral truth, separate social communities derive their own moral narrative, 'oriented to a distinctive perspective, heritage and vision of life' (MacIntyre 1993: 216).[13] Whilst many would question the extent to which this moral relativism holds true (given the prevalence of some fundamental and generally accepted precepts regarding human rights), the selectivity and interpretation of moral principles in the absence of a unifying source is self-evident and it is clear that what is approved of, or at least tolerated, in one culture is not necessarily in the next.[14] As a profession, the military per se may be seen as a distinct moral community with culturally defined moral reasoning derived from its particular relationship to violence, and that finds its profession-wide expression in pertaining international law and convention. But, this is a quasi-autonomy, subject to the profession's relationship to society and the state, from which it derives its existence, sustenance and purpose. Members of a profession are awarded only 'limited professional exemption' from societal norms in order to perform their expert functions, their commitment to serve society's values providing the 'external' or 'transcendent' moral reference, diverting the military community from self-justification (Torrance 1998: 26). Identifying a coherent, pervasive and consistent societal or national 'value set' may be difficult, given its temporal nature and the plurality of cultures, attitudes and beliefs to be found within most modern states.[15] However, a level of tacit social agreement may be articulated in constitutions, domestic law and policy, and indicated in social acceptance or aversion to certain behaviour or concepts, identified (with significant caveats) in polls, elections and through the national media. The military's moral divergence from these societal norms and values, where it exists, needs to be clearly articulated in terms of the requirements of performing expert functions to expedite the professional tasks set. Hence, the moral role/identity of a member of a professional military is based to a large extent on the relationship between the values and interests of the society that it represents (and from which it is drawn) and the functional requirement of service.

Specific military 'Ethics' must be understood, however, not only in terms of the supporting societies from which individual military personnel are drawn, but also in terms of the particular military institutions within which they are adopted, internalised, and entrenched. Each of these disparate institutions develops its own sub-culture, which embodies the 'values, norms, and assumptions' that predispose its members to 'interpret situations in a limited number of ways' (Wilson 2008).[16] This culture informs the identity of the military community, and is promoted through the structures of authority and reward, through cultural norms (its values and standards), and through the ethos (the spirit of the organisation) that condition its existence.

Consideration of the institutional form adopted by various militaries, including their structure, declared precepts and formalised processes, will offer some insight into the prevailing surface culture or climate of the establishment. All militaries tend towards highly formalized hierarchical structures broken down into subordinate units, with centralized control from the top. The prevalence of hierarchical forms may undermine the concept of the military as a profession, (a profession being defined in part by its peer-oriented structure, the professional autonomy of its members and its self-reflective nature), but the rigidity of these bureaucratic structures may vary and power and authority, communication, and leadership may be exercised in both command style and in more subtle forms. These overt processes, structures and symbols, including the doctrinal body that underpins them, have been referred to by Schein (2004: 25–27) as artefacts and represent a superficial level of an organisation's culture that, nonetheless, contributes to the socialisation of its members.

Socialisation into military institutions is supported by expressed values and standards, an articulation of those things seen as worthy by the particular military community and a conscious expression of the desired behaviour of its members. Understanding these core values is central to a sense of organisational (and professional) identity. Expressed in codes or value lists that differ dependent on nationality, service and even unit, these declared values nevertheless display a surprising degree of similarity across the military profession, including courage, loyalty, discipline, self-less commitment, respect and integrity. As expressions of desirable characteristics or virtues rather that values (those things held dear) per se, they identify those traits or tendencies that contribute to the effective pursuit of established aims. As such, they may be seen as purely instrumental or functional codes that may be practised by the unjust as much as by the just: they are not in themselves good, but only take on moral worth in the context of the good that they serve. It is the professional Ethic that unites the functional with the moral, associating what are valued behaviours with their moral objective. The utility of professional and organisational values as a guide to right behaviours requires clarity of meaning and reflection on application. Where values are not well understood by serving personnel,[17] neither the moral nor instrumental significance of, for example, discipline or obedience may be evident, whilst honour and loyalty might be confused with group attachment, becoming counter-values that undermine professional integrity (Coleman 2013: 45–51).

Knowledge of organisational or professions values and the principles derived from them is not in itself sufficient to ensure right behaviours.[18] If the PME is to both inform and reflect the meaning and identity of the institution, it must be at the core of its deep culture, that is, at the level of shared assumption of its members, the implicit and unconscious understandings that collectively interlock and represent the cultural paradigm of the group (English 2004: 15–18). These assumptions, that shape individual and group behaviours within the community, are influenced by the experience of past successes, and perpetuated through self-imaging appointment and promotion. They are translated through explicit and implicit rules, encouraged by a collective narrative of common experience and promoted through role models, symbols, ritualised behaviours and expressions of oneness that often persist as powerful symbols long after the reasons for their adoption have been lost. (Snider *et al.* 2009: 6). These assumptions, and the practices that support them, may be conducive to the commitment of the individual to their institution, and a contribution to the efficacy and ethicality of their service. Deep engagement with, and immersion in, the moral community provides soldiers with necessary decision-making 'tools' drawn from their professional identity. As an example, highlighted to the author by a trainer at their Lympstone training establishment, if a Royal Marine precedes the question 'what should I do' with 'what should a Marine do', and defines this in professional terms that have personal resonance, then they are much nearer to answering the question. This professional identity is the 'being' element of service. Where the artefacts, values, and assumptions of the moral community are grounded in the wider concept of a PME, they have coherence, applicability and purpose and, where identity is strong, provide an important form of moral guidance and social control.

An awareness of moral community and of the prevalence of ingrained cultural preferences, supported by tradition and practices, may help to overcome the institutional inertia that is so often confused with reason, but which challenges functionally as well as ethically required change. Consider, for example, the hue and cry from traditionalists regarding the recent and well-overdue reform of the Royal Navy toast – 'to wives and sweethearts', with the aside 'may they never meet' – an affront to socially accepted standards and hardly conducive to the engendering of an inclusive and effective working service environment. A moral community, and the culture that sustains it, can equally be self-indulgent and even cult-like, where the ethics of professional service are lost or distorted. Inter- and intra-service parochialism may contribute to the emergence of exclusive sub-cultures as services and specialism seek greater recognition, freedom of action, status and resources, encouraging exaggerated pronouncements, doctrines and strategies and an emphasis on difference that may undermine broader commitments to the health of the military profession (or at least its national representation) as a whole. As one requirement of a coherent, relevant and applicable

PME, addressing the nature of the moral community, through reflection on its practices, values and assumptions, is an essential contribution to the development and maintenance of a 'current' and professional military service, sure of its moral identity and reflective of its moral calling, purpose and expertise.

The organic PME and the educational challenge

The framework for defining a PME outlined above recognises both a universal professional coherence and an organic particularism embedded in the concept, which is both widely applicable and subject to local conditions and interpretation. Articulated as values and rules, and in principles that are 'less vague than values and less specific than rules' (Imiola 2010: 17) the PME invites judgement in application consistent with the definition of professionalism. Collectively, these sources of ethical guidance reflect the functional requirements of the service in order to have meaning to the professional community, accessible to the military at all levels and for all actions. Underpinning morals may provide a foundation for professional, as well as personal behaviours, but the interpretation of the Ethic will be influenced by the context of the military force, differing one from another as it reflects the context and culture of a service. Drawing on both universal and localised understandings of professional morality as it applies to military activity, the particular Ethic of national Armed Forces, Service organisations or units, will be defined according to the four core and interlinking moral concepts outlined above, perhaps most accessibly articulated as 'who is served', 'what is done', 'what is known', and 'who we are'. A clear understanding of these four areas of the Ethic, interpreted and defined within the specific institutional context, will provide the basis for ethical education and development applicable to all members of the profession.

This level of particularism poses some self-evident challenges for the development of a PME, and certainly for any enduring and universal Ethic that might be applied across the various military organisations that constitute the profession. But this, in itself, is a useful starting point for the discussion of the 'what, how and why' of the educational task. At the heart of any ethical programme of education must be an understanding of the conceptual ethical basis of the profession, as defined and promoted by its particular institutional representation. Most clearly this suggests the requirement to resist the 'one size fits all' assumptions of much of the current discourse on Professional Military Ethics Education (PMEE) in favour of a more nuanced and culturally sensitive approach, based on local 'dialect' and grounded in the military's ethical professionalism.

Additionally, this approach to the PME, as a basis for informing ethics education, challenges the overstated distinction between functional and aspirational training and education. By situating the Ethic in its professional context, the relationship between the moral concepts and the functional requirement is at the forefront of ethical discourse and understanding. The separation between Professional and Character Development – the former being essentially functionally instrumental, the latter having 'moral purpose' (Wolfendale 2008: 164) – is a false divide. The distinction between professional 'rules' for good functioning and aspirational values is evidently not so clear-cut. From the perspective of a PME defined in terms of the four moral concepts of authority, purpose, expertise and community, the military, as a profession, is recognised as both moral and functional, and the distinction between the two is, at least in terms of effect, a less relevant one. It is a matter for philosophers to determine whether a soldier is to be considered less moral because he behaves in a manner consistent with moral principles, but is motivated by his professional identity derived in large part from the Ethic of his profession. It would be a missed opportunity, however, if the ethical purist's requirement for 'goodness for goodness sake' denied the benefits of more indirect and potentially more effective means of deriving morally consistent outcomes. Where making sound moral decisions is recognized as an essential part of professional efficacy; and where the question, 'what are my professional duties and operational aims?' is accompanied by 'what is it right for me as a professional soldier, sailor or airman to do?'; the moral autonomy of the individual professional need not be compromised. Of course, there are examples of desensitisation, indoctrination and the by-passing of individual moral thinking to be found in training, and this is an area of challenge where the professional requirement to kill is at odds with the moral precepts of right to life and against murder or physical attack. But the satisfaction of military purpose also requires individuals able to make difficult moral judgements in complex environments, to be capable of facilitating mission command based on an understanding of intent and the values that underpin it, and to be able to take individual responsibility for their actions in the light of their moral expertise. The PME should provide the foundation for reflection on the moral principles of the profession at large and of the institutions that the profession inhabits. Where PMEE can offer the members of the profession a sound understanding of the four conceptual elements that constitute the Ethic, individual servicemen and women will be far better equipped to perform their professional duties in both a functionally and ethically enlightened manner.

Notes

1 Given the military's penchant for acronyms, these four elements might be collectively termed PACE (Purpose, Authority, Community, Expertise).

2 In recent operations, this has proven to be the case despite attitudes towards the rightness or success of the deployment, or even, contentiously, the behaviour of individual service personnel. In the case of Royal Marine Sergeant Alexander Blackman (known as 'Marine A'), convicted in 2013 of the murder of a gravely wounded and unarmed Afghan insurgent in 2011, over 100,000 people signed an online petition demanding his release and stating that 'a soldier should never go to prison for killing the enemy in a battlefield situation'. Blackman was released in April 2017, the Court of Appeal reducing his conviction to manslaughter on the grounds of diminished responsibility after reviewing evidence that he had been suffering from stress related 'adjustment disorder' at the time of the killing.

3 See, for example, the catalogue of misconduct by senior military commanders in the US Armed Forces reported in the *Washington Post* in 2014 (Whitlock 2014).

4 For example, following the findings of the Baha Mousa inquiry, the November 2009 Watts–Andrews inquiry and the Charles Kirke study of January 2011, the British Army initiated a project to overcome the low level of understanding of values and standards amongst the NCO cohort.

5 The 1948 *Universal Declaration of Human Rights* is a representation of commonly accepted foundational rights, whilst the tenets of the Just War tradition have been widely accepted as the ethical basis for legitimate military action.

6 For a medical professional, for example, this prescribed 'good' would be health, for a legal professional, justice.

7 These include its contribution to the satisfaction of higher strategic interests, such as the maintenance of domestic and international support for operations necessary for the successful achievement of objectives, the operational and tactical benefits associated with winning 'hearts and minds' and of maintaining the morale and psychological health of the soldier.

8 Consider, for example, the North Korean People's Army, with its constituent Korean People's Army Ground Force and Air Force and Korean People's Navy.

9 Territorial defence remains a military priority in the UK's 2010 *Strategic Defence and Security Review*, although the *National Security Strategy* from which it was derived identifies a state-based assault on sovereign territory as a tier three (lowest priority) risk, suggesting that other activities are more likely to occupy British forces in the coming years.

10 Of course, killing and dying may have little to do with current military activities (reconstruction, training, etc.), and has been of less relevance to some branches of the Services than others, whilst the remoteness of new technologies raises new ethical challenges regarding the moral status of risk-free 'drive-by' killing (Leveringhaus and Giacca 2014).

11 For example, see *US Army's Field Manual 3–24*, section 13–6, which sets out 'ten basic rules'.

12 A recording played during the prosecution of 'Marine A', identifies Blackman's words to his victim – 'There you are. Shuffle off this mortal coil, you c***. It's nothing you wouldn't do to us' – before turning to his comrades and saying 'Obviously this doesn't go anywhere, fellas. I just broke the Geneva Convention'.

13 Nagel (2006: 653–659) argues that 'due to the lack of a commonly accepted theory of the morality of war, ethical views are embedded in society and will evolve through time shaped by context and common experience'.

14 Consider attitudes toward the morality of abortion, polygamy, capital punishment, homosexuality and slavery, all of which have been time and context specific.

15 British Prime Minister David Cameron, writing in the *Mail on Sunday* (15 June 2014) to mark the 799th anniversary of the Magna Carta, stated that:

> a belief in freedom, tolerance of others, accepting personal and social responsibility, respecting and upholding the rule of law – are the things we should try to live by every day. To me they're as British as the Union Flag, as football, as fish and chips.

16 In the US Army's revised *Profession of Arms* paper, (US Army 2010: 2–3) the Army Ethic is defined as:

> the collection of values, beliefs, ideals, and principles held by the Army Profession and embedded in its culture that are taught to, internalized by, and practiced by its members to guide the ethical conduct of the Army in defence of and service to the Nation.

17 Interviews with Royal Naval officer cadets and mid-career officers over 2012–14 suggested that, whilst most were aware of their Service's core values, articulated as C2DRIL: Commitment, Courage, Discipline, Respect, Integrity and Loyalty, there was only a poor understanding of their ethical significance.

18 For example, despite service assumptions of shared and understood values, the 2006 report by the Mental Health Advisory Team for operation Iraqi Freedom, found that only 47 per cent of soldiers and 38 per cent of Marines surveyed agreed that non-combatants should be treated with dignity and respect, 41 per cent and 44 per cent respectively agreeing that torture should be allowed to gather important information where it would save the life of a comrade (MHAT IV 2006).

Bibliography

Bell, D.M. (2012) 'The Moral Crisis of Just War: Beyond Deontology Towards a Professional Military Ethic', *Journal of Faith and War* (2 July), online at http://faithandwar.org.

British Army (2008) *The Aitken Report. An Investigation into Cases of Deliberate Abuse and Unlawful Killing in Iraq in 2003 and 2004.*

British Army, Director Leadership, *The Army Leadership Code. An Introductory Guide,* September 2015, online at www.army.mod.uk/documents/general/rmas_AC72021-TheArmyLeadershipCode.pdf#search=ValuesandStandards.

Burk, J. (2002) 'Theories of Democratic Civil-Military Relations', *Armed Forces and Society* 29:1, 7–29.

Burk, J. (2005) 'Expertise, Jurisdiction and Legitimacy', in D.M. Snider and L.J. Matthews (eds.), *The Future of the Army Profession*, 2nd edition, New York: McGraw-Hill, 39–50.

Carrick, D. (2008) 'The Future of Ethics Education in the Military', in P. Robinson, N. de Lee and D. Carrick (eds.), *Ethics Education in the Military*, Aldershot: Ashgate.

Case, C., Underwood, B. and Hannah, S.T. (2010) 'Owning our Army Ethic', *Military Review*, September, 3–10.

Casey, G.W. (2009) 'Advancing the Army Professional Military Ethic', *Joint Forces Quarterly*, 54(3), 14–15.

Chadha, V. (2013) 'Role of Morals, Ethics and Motivation in a Counter-insurgency Environment', *Journal of Defence Studies*, 7(2), 49–68.

Coleman, S. (2013), *Military Ethics. An Introduction with Case Studies*, New York: Oxford University Press.

Cook, M.L. (2004) *The Moral Warrior: Ethics and the Service in the US Military*, New York: State University of New York Press.

Cook, M.L. (2013) *Issues in Military Ethics. To Support and Defend the Constitution,* Albany: State University of New York Press.

Davis, M. (2003) 'What can we Learn by Looking for the First Code of Professional Ethics?', *Theoretical Medicine and Bioethics*, 24(5), 433–454.

Dobbin, V. (2010) 'Ethics and the Military Community', *Journal of Defense Resources Management*, 1(1), 69–76.

Economist (2013) 'Liberty's Lost Decade', *The Economist*, 3 August, 11.

English, A.D. (2004) *Understanding Military Culture. A Canadian Perspective*, Montreal: McGill-Queen's University Press.

Gulhane, J. and Taha, R.M. (2013) 'Presidential Decree Details Military Oath', online at www.dailynewsegypt.com/2013/08/28/presidential-decree-details-military-oath/.

Hartle, A.E. (2004) *Moral Issues in Military Decision Making*, 2nd edition, Kansas: University Press of Kansas.

Haynes, D. and Pitel, L. (2012) 'Ex-officers Face Whitehall Ban over Lobbying Claims', *The Times*, 15 October.

Hines, L.A., Gribble, R., Wessely, S., Dandeker, C. and Fear, N.T. (2014) 'Are the Armed Forces Understood and Supported by the Public? A View from the United Kingdom', *Armed Forces and Society*, 1–26.

Houghton, General Sir N. (2013) Annual Chief of Defence Staff Lecture, RUSI, White-hall, London, 18 December, online at https://rusi.org/event/annual-chief-defence-staff-lecture-2013.

Imiola, B. and Cazier, D. (2010) 'On the Road to Articulating Our Professional Ethic', *Military Review*, September, 11–18.

Leveringhaus, A. and Giacca, G. (2014) *Robo-Wars: The Regulation of Robotic Weapons*, Oxford Martin Policy Papers 2, online at www.oxfordmartin.ox.ac.uk/downloads/briefings/Robo-Wars.pdf.

Lucas, G. Jr., (2003) 'From *Jus Ad Bellum* to *Jus Ad Pacem*: Re-Thinking Just-War Criteria for the Use of Military Force for Humanitarian Ends', in D.K Chatterjee and D.E Scheid (eds.), *Ethics and Foreign Intervention*, Cambridge: Cambridge University Press, 72–96.

MacIntyre, A. (1993) *After Virtue*, 2nd edition, Guildford: Biddles Ltd.

Moten, M. (2010) *The Army Officer's Professional Ethic – Past, Present, Future*, Carlisle, PA: Strategic Studies Institute, US Army War College.

Nagel, T. (2006) 'The Logic of Hostility', in G. Reichberg, H. Syse and E. Begby (eds.), *The Ethics of War. Classic and Contemporary Readings*, Oxford: Blackwell, 653–659.

Office of the Surgeon, Multinational Force-Iraq and Office of The Surgeon-General, United States Army Medical Command, *Mental Health Advisory Team (MHAT) IV. Operation Iraqi Freedom 05-07, Final Report*, 17 November 2006, online at www.combatreform.org/MHAT_IV_Report_17NOV06.pdf.

Olsthorn, P. (2011) *Military Ethics and Virtues. An Interdisciplinary Approach for the 21st Century*, London: Routledge.

Schein, E.H. (2004) *Organizational Culture and Leadership*, 3rd edition, San Francisco: Jossey-Bass.

Senge, P. (2006) *The Fifth Discipline: The Art and Practice of the Learning Organization*, 2nd edition, London: Random House.

Snider, D., Oh, P. and Toner, K. (2009) *The Army's Professional Military Ethic in an Era of Persistent Conflict*, Carlisle, PA: Strategic Studies Institute, US Army War College.

St George, A. (2011) *Royal Navy Way of Leadership*, London: Preface.

Taylor, M.D. (1978) 'A Professional Ethic for the Military', *Army*, 18–21 May.

Toner, J.H. (1995) *True Faith and Allegiance: The Burden of Military Ethics*, Lexington: University of Kentucky.

Torrance, I. (1998) *Ethics and the Military Community*, Strategic and Combat Studies Institute, Occasional Paper 34.

Waters, C.P.M. (2011) 'War Law and its Intersections', in D. Whetham (ed.), *Ethics, Law and Military Operations*, Basingstoke: Palgrave Macmillan.

Whitlock, C. (2014) 'Military Brass, Behaving Badly: Files Detail a Spate of Misconduct Dogging Armed Forces', *Washington Post*, 26 January, online at www.washingtonpost.com/world/national-security/military-brass-behaving-badly-files-detail-a-spate-of-misconduct-dogging-armed-forces/2014/01/26/4d06c770-843d-11e3-bbe5-6a2a3141e3a9_story.html?utm_term=.d91e73844128.

Wolfendale, J. (2008) 'What Is the Point of Teaching Ethics in the Military', in P. Robinson, N. de Lee and D. Carrick (eds.), *Ethics Education in the Military*, Aldershot: Ashgate, 161–174.

Wolfendale, J. (2009) 'Preventing Torture in Counter-insurgency Operations', in D. Carrick, J. Connelly, and P. Robinson (eds.), *Ethics Education for Irregular Warfare*, Farnham: Ashgate, 57–74.

Wilson, P.H. (2008) 'Defining Military Culture', *The Journal of Military History*, 72(1), 11–41.

US Army (2010) *Army: Profession of Arms*, Centre for the Army Profession and Ethic.

US Army (2013) *The Army* Profession, ADRP 1, 14 June, Washington, DC: HQ, Department of the Army.

US Army (2014) FM 3-24/MCWP 3-33.5, *Insurgencies and Countering Insurgencies*, Washington, DC: HQ, Department of the Army.

7 Ethical challenges for the modern military

John Thomas

Ethical dilemmas for individual combatants have, to a greater or lesser extent, been a consequence of warfare since the formulation of the Just War Theory. Nevertheless, it would be true to say that in many historical conflicts the ethical rules of war, such as proportionality and fair treatment of prisoners, were often ignored in favour of a more utilitarian and brutal emphasis on success at all costs. However, especially since the Nuremberg tribunals which followed the end of the Second World War, the principle that members of the military can be held individually responsible for their actions has taken root; 'just following orders' is now a particularly lame and unacceptable justification for unethical behaviour.

The paradigm of armies fighting armies is now much less common, although it is unlikely that it has disappeared forever. Over the past 60 years or so, it has been supplemented by a number of subtypes, including international terrorism and humanitarian intervention, in which the modern soldier, sailor or airman is confronted by a more complex combat environment. Additionally, he or she is increasingly under a media spotlight, whether it be by traditional electronic or print media, citizen journalism, or even self-imposed through use of their own mobile devices. The modern combatant can therefore face a new double whammy of increased complexity and increased scrutiny. These factors, coupled with the long duration of modern interventions which can now involve repeat tours of duty, can impose great stresses on individuals.

In this chapter, I will look at some of the ways in which modern warfare challenges the ethical well-being of the individual combatant. I shall remind readers of the true nature of war, then, by using some real-world examples of ethical dilemmas that have been faced in recent conflicts, demonstrate how difficult and even tragic these dilemmas can be. Finally, I would like to suggest how commanders might begin to deal with this issue for themselves and those they command.

Even with 24-hour news coverage and sometimes uncensored internet footage, most people view war through a sanitising filter. It is something that arouses our concern, even disgust, but our distance from the actual horrors means that they do not threaten to overwhelm us with their sights, sounds and smells. The following short passage gives an idea of these all-encompassing horrors:

Those who claim to love war must have been involved far from the carnage of the battlefield, corpses scattered randomly, disembowelled women. War is an absolute evil. There is no joyful war, or sad war, beautiful war or dirty war. War is blood, suffering, burnt faces, eyes widened by fever, rain, mud, excrement, filth, rats running over bodies, monstrous wounds, women and children turned into carrion. War humiliates, dishonours, degrades. It's the world's horror brought together in a paroxysm of squalor, blood, tears, sweat and urine.

(Royal 2012: 9)

This brief passage graphically illustrates the most enduring ethical challenge for the military. War is an activity in which the utterly abnormal becomes distressingly normal, in which human reactions and emotions can be numbed by repeated exposure to scenes of utter horror. Men and women can be enraged by the deaths of comrades and family, and become fuelled by hatred and revenge. The physical pressures of tiredness, hunger, heat, cold and fear can exaggerate all these negative reactions. This is not a new problem: its traumatic effects remain as powerful as ever. Those who fought wars in previous decades, or even previous centuries, were exposed to similar pressures. But, as we shall see, the changing nature of warfare has added some new difficulties to the many that already existed.

Because of the horrors they will encounter, those who engage in war must expect their own ethical view of the world to be severely challenged. Is the war a just one? Have they been able to find the right balance between mission success, protecting innocent non-combatants, respecting the laws of war and ensuring their own troops are not placed at unnecessary risk? Why must they respect ethical standards that seem to hand the advantage to my enemy? In the cauldron of war, these are hard questions, and the answers are often unsatisfactory and elusive.

However, once a decision has been taken to commit to armed action, that action must be pursued unflinchingly and in the most effective way. Behaving ethically should not be confused with lack of decisiveness, or a lack of resolution. The difficulty for the modern military is how to maintain a sense of perspective and, for commanders, to lead those under them to ensure an honourable and durable victory. Furthermore, an ethical approach is not the same as one from the point of view of fairness, in the sense that both sides must be equal to ensure a 'fair' trial of strength. Although there are rules which govern the conduct of war, it is not a game in which both sides have to be equal. For example, the lack of an enemy air defence system does not preclude me from using my air force against that enemy. If my enemy does not provide its forces with adequate clothing for cold weather operations (as Argentina failed to do with its conscripts in the Falklands War), that is no reason for me to reduce the effectiveness of my troops by doing likewise. But even though a conflict can contain thousands of examples of such unfairness, it can still be just and fought ethically.

There are many ethical challenges which face the modern military, and space does not permit me to look at them all. There are new challenges which result from technology, such as the use of unmanned aerial vehicles, and from the society from which members of the armed forces are recruited. Many of those joining the armed forces come from a world of moral relativism rather than one of ethical rigour. We also find that new ethical challenges for both the military and society are emerging as the long-term effects of life-changing injuries, both physical and mental, demand long-term care, often extending for many years beyond discharge.

I will therefore concentrate mainly on the ethical challenges which arise from conflicts that do not take place between states; for example, 'revolutionary war', the 'war on international terrorism', peace enforcement, humanitarian intervention and so on.

Taking a template outlined by Christopher Coker in his book *Ethics and War in the 21st Century* (Coker 2008: 62) as a basis, we can see clearly how the character of war began to change in the second part of the twentieth century, by looking at his comparison between the Second World War and the Algerian war of independence, which began in 1954:

- During the Second World War, combat generally took place at a specific moment and place – a battlefield – whereas in the Algerian war the theatre of operations was indeterminate, that is to say combat could occur anywhere.
- In the Second World War, it was also generally possible to say when an engagement had begun and ended, whereas in the Algerian war, there was a multiplicity of simultaneous engagements.
- In the Second World War, the actors were well defined – soldiers who were members of armies, whereas in Algeria it was difficult, and sometimes impossible, to distinguish between combatants and civilians.
- Similarly, there was a difference between the way armies attacked armies in the Second World War, and the way that distinction between civilian and military targets became blurred in Algeria.
- And finally, the Second World War saw the use of traditional weaponry as opposed to the increasing use of asymmetric warfare in Algeria.

The nature of modern warfare has continued to evolve continuously since the Algerian war. After the 'revolutionary war' in Algeria came Vietnam, arguably the first major asymmetric war, where very large numbers of low-tech foot soldiers confronted ostensibly overwhelming, technologically driven, firepower. Then came peacekeeping, peace enforcement and what I refer to as 'applied humanitarian intervention',[1] which to a large extent overturned the longstanding presumption of state inviolability. International terrorism, spawned by domestic terrorism but developed far beyond that, grew from small beginnings until it reached an initial dreadful climax on 9/11. We have recently seen how the group known as ISIL or ISIS has been able to become the first terrorist group in modern times to take and hold large swathes of territory and style itself as a 'state'. ISIL also poses a particular dilemma because it sees its actions as justified on theological grounds, untempered by ethical or legal considerations.

As the British general, Rupert Smith said in his book *The Utility of Force: The Art of War in the Modern World* (Smith 2005: 267ff.), we are now in an era of 'wars amongst the people'. The paradigm is no longer only the relatively straightforward matter of defeating an enemy army, although that might still be part of what is necessary. The strategic objective is increasingly the will of the civilian population. This is radically different from the Second World War when the allied armies defeated the German armed forces. At that stage, no one cared about the will of the German people: they had no option but to tolerate defeat (although many of them may have felt that, along with most of Europe, they too had been liberated from Nazism).

If the objective is indeed to win the will of the people, then these new circumstances have important consequences for the need for troops to behave ethically. The strategic aim of winning the will of a people, who may be suspicious or even initially hostile, will be seriously undermined by instances of unethical behaviour. In such circumstances, relatively isolated incidents can assume strategic significance.

For example, there has been no significant contention about the unknown number of Iraqi armed forces killed by British forces during the combat phase of the second Gulf war. However, the death of a single young man, Baha Mousa, who died while in the custody of British soldiers and after having been assaulted by a number of them, not only provoked widespread hostility in Iraq, but almost overshadowed the bravery and achievements of thousands of other UK forces. It also led to extensive negative media coverage for the British Army in particular and to a lengthy public inquiry in the UK.[2] Actions such as those which led to the death of Baha Mousa, or the even more widely known Abu Ghraib mistreatment of prisoners, are precisely how wars amongst the people can be lost.

Winning the war, or as it might now be termed 'the kinetic phase', is no longer the military's only task. They now have to win the peace as well. This imposes real challenges for the armed forces. Before deployment, they have to be trained to a very high pitch of combat readiness, and psychologically prepared to kill or be killed. However, a short combat phase can quickly be followed by long occupation and reconciliation/reconstruction phases. During these phases, there can nevertheless be recurring outbreaks of deadly insurgent or terrorist activity. Not only can it be extremely difficult for the military to know whether they are supposed to be war fighters, peacekeepers, surrogate police or even civil engineers, but this confused and confusing physical and psychological situation serves also to complicate the ethical landscape. What is the relationship of the military with the indigenous population? Are they victors with total authority over them, or are they a humanitarian force acting in partnership with a new government? To what extent can overwhelming military force be used to quell subsequent insurgencies once 'peace' has been declared? The situation will vary from conflict to conflict and from time to time, but each of these different possibilities implies different legal and ethical underpinnings to the mission.

I now give some examples of some of the challenges that these new types of war pose, using a number of real-world examples; the first is an example which I have summarised from General Royal's book.[3]

It concerns two French navy fighter crews called in to support coalition ground forces under attack in Afghanistan. The ground controller has asked them to bomb a farm compound, from where insurgents are firing on coalition forces. Time is of the essence on two counts; first, because the situation on the ground is desperate, and second because the aircrafts' fuel is running low. Using their laser-guided bombs, the crews can almost guarantee to destroy the compound they fire at. But they can see three compounds, not one, and cannot tell from 15,000 feet from which compound the firing is coming, nor can they tell if innocent civilians are taking cover in the other two. The crews decide that they have no alternative but to turn for home without releasing their weapons 'with a terrible feeling of impotence, of disappointment and frustration, but with the conviction that, on that day, there were no acceptable alternatives'.[4]

In this example, we see an age-old dilemma – whether or not to open fire when non-combatants might be present – not eased, but made more acute by imperfect technology. In years gone by, aircraft crews would have had to trust the information given to them either by the controller on the ground or during a pre-flight briefing. Now, thanks to the quality of modern optics, an aircraft crew can see just enough to introduce ethical uncertainty into their decision making. The technology at their disposal was clearly of benefit to any non-combatants who might (or might not) have been present on the ground, but it was also of benefit to the insurgents, and clearly not of benefit to coalition ground forces. Is it therefore any wonder that the crews had 'a terrible feeling of impotence, of disappointment and frustration'?[5] The crew's reaction also highlights that difficult ethical decisions can carry long shadows of regret and doubt about whether or not the right decision was made.

In Afghanistan, the Taliban often used young and unarmed youths to report on the movements of British and other ground patrols. These lookouts reported the movements of International Security Assistance Force (ISAF) troops by mobile phone. And this information was frequently used by the Taliban to lay Improvised Explosive Devices (IEDs), or to set ambushes targeting coalition forces. So, were these young, ostensibly civilian and unarmed lookouts legitimate targets or not? If you were a young NCO leading a patrol that you felt was being led into mortal danger, how would you answer that question?[6]

Many would say that, even though these lookouts might be young and ostensibly civilians, they were providing direct help to the enemy and that they should always have been considered a legitimate target. However, what about those who laundered Taliban money or provided them with safe accommodation, or storage for their weapons? How far from the front line and the enemy holding a weapon does legitimacy extend in a war amongst the people?

The next example was common to the Balkans and Afghanistan wars. In the Balkans, much of the fighting was carried out by irregular militias drawn from the various ethnic groupings that were at each others' throats. Under the rules of engagement, if a member of a militia picked up a weapon and fired at you, then he became a legitimate target. If he put the weapon down and walked away, he instantly became a civilian again and no longer a legitimate target. If you passed this man in the street the following day, how would you have reacted? This scenario was repeated in Afghanistan in 2012 when a change in the Rules of Engagement (ROE) allowed coalition troops to open fire only if they were being attacked or were in imminent danger. The Taliban exploited this change and became adept at hiding weapons in prepared firing positions, knowing that they were immune from engagement unless and until they picked them up. Some British troops at the time felt that they had become little more than bait to draw the enemy into the open.

Thus, in addition to all the pre-existing difficulties inherent in the Afghan campaign, a new level of ethical complexity – or confusion – was added. This was a 'hot war', in which the Taliban felt no compunction to observe the normal conventions or laws of war. Many coalition forces felt that their own governments valued the lives of those actively engaged in trying to kill them above their own. Such erosion of confidence in the political leadership can be very corrosive, as it can undermine individual soldiers' willingness to respect strict ethical and legal guidelines which are felt to have been imposed by political leadership that was ignorant of the true stresses of war.

As part of the same trend, there has been a move to make Service personnel on combat duty subject to legislation and levels of scrutiny designed for civilians in peacetime. There is an unresolved tension between International Humanitarian Law and the Laws of Armed Conflict and human rights law. On the one hand is the ability of foreign nationals to lodge claims based on the extraterritorial application of the European Convention of Human Rights against the UK MOD for (potentially) every death in an armed conflict imposes a theoretically unlimited potential for retrospective claims. On the other hand, is the failure of the principle of combat immunity, thereby allowing families and individuals to bring negligence cases against the MOD for injury or death[7] risks the unwitting application of peacetime Health and Safety norms to warfare.

A third example is also from the Balkans war, and highlights the problems that can arise for a neutral peacekeeping force. A group of women and children was taking refuge in house in a town that was under siege and likely to be overrun very soon. Saving those people from near certain death would seem to be an obvious act of humanity, but it would also be seen a 'taking sides' in a conflict where the neutrality of the peacekeeping force was its main strategic asset.

Ethical problems are, of course, not confined to the tactical level of warfare. Take the hypothetical case where a battalion commander knows that his troops do not have enough blast resistant armoured vehicles to allow them all to patrol in relative safety. In order to complete their mission, some of them will have to patrol in soft-skinned vehicles, making them very vulnerable to small arms fire and particularly to roadside IEDs. Indeed, some of these troops may have already died or been badly injured in such circumstances. To which would you give priority: the mission objectives or the safety of your own troops? And how would you arrive at that decision and subsequently explain it, not to the media, or to your commander, but to the widow and children of a soldier who subsequently lost his life in a roadside bomb attack on a soft skinned vehicle? This widow will not be interested in the economics or politics of your decision, but in the quality of your ethical decision making (and you might also find yourself exposed to prosecution for the reasons outlined above).

Although presented here as a hypothetical case, these were exactly the circumstances faced by British commanders in the second Gulf War, when:

> The [UK] Ministry of Defence (MOD) was slow in responding to the developing threat in Iraq from IEDs. The range of protected mobility options available to commanders in MND(SE) was limited. Although work had begun before 2002 to source an additional PPV, it was only ordered in July 2006 following Ministerial intervention.[8]

An example of a strategic dilemma from the Second World War now follows: it vividly illustrates the sometimes multiple and delicate ethical balances that have to be struck.

Reinhardt Heydrich was a Nazi SS chief who was appointed by Hitler as 'Reich Protector' to Czechoslovakia in 1941. His main mission was to put an end to Czech resistance, which had become more violent and widespread following the German invasion of the Soviet Union. Amongst other actions, Heydrich ordered that the number of executions and deportations of those resisting Nazi occupation should be increased significantly.

As a serving military officer during a war, there was little doubt that Heydrich was a legitimate target. If he were killed, he would be mourned by very few. However, his assassination would also be likely to provoke a savage reaction by Hitler, who would probably seek to avenge his death by killing thousands of innocent Czechs. Would Heydrich's death be worth it?

For some, however, there was an even bigger issue at stake than the potential deaths of many innocent Czechs. In London, where the Czech government was in exile, Czech resistance was seen as ineffective. As a result, the Czech case was not being heard (unlike, for example, that of the Poles). There was therefore a great risk that the allies would simply give up on Czechoslovakia and accept the country's much reduced frontiers, as agreed at Munich in 1938 by Germany, Italy, France and the United Kingdom. (The Czechs themselves were not party to these negotiations, but were presented with a *fait accompli*).

Overall, it would not be an exaggeration to say that the future of an entire country was in play.

The Czech resistance was vociferously opposed to the assassination, fearing not only reprisals but the decimation of their own networks. However, Benes, the Czech prime minister in exile, was of the view that so high were the stakes that 'even great sacrifices would be worth it' (Burleigh, 2011: 306).

The assassination, undertaken by Czechs trained by the British Special Operations Executive, went ahead. Heydrich was killed and the reprisals were indeed severe. One village, Lidice, which was believed by the Germans to have had some involvement in the plot, was burned to the ground, all the men killed and the women sent to concentration camps; 82 children were also killed. The ruins of the village were blown up and even the gravestones removed to be used as building materials. In addition, and as feared, the entire underground resistance was captured, with 3,188 arrests and 1,357 sentenced to death.

However, as a result of the operation, the British renounced the Munich agreement, meaning that after the war the Czechs would recover the Sudetenland and regain their old borders; so, in that strategic sense the country was saved. The reprisals also helped to galvanise the allies against the Nazis and increased their moral capital. But the moral calculus was, to say the least, a difficult one, especially as many of those it was feared would be (and were) killed were civilian non-combatants, including children.[9]

The Heydrich assassination is instructive, not merely because it weaves together ethical, political and humanitarian dilemmas, but also because the increased feasibility of targeted assassinations in our own era – especially using Unmanned Aerial Vehicles (UAVs, commonly called drones) – raises many of the same issues.

In an era where there is much talk of the 'Strategic Corporal' it is as well that we remind ourselves from time to time that however real are the ethical difficulties on the front line, those posed in the more rarefied political atmosphere of Whitehall or the White House can be every bit as challenging.

Ethical dilemmas can also confront those who hold the very highest military appointments. In 2003, as war with Iraq appeared ever more likely, debate raged in the United Nations (UN) about the legality of any potential invasion. The more the diplomats and politicians spoke, the less clear the answer to the central question of legality became. In the UK, Lord Goldsmith, the government's senior law officer, had briefed Prime Minister Tony Blair in 2002 that a new UN Security Council resolution would be needed to authorise military action; in Goldsmith's view, resolutions 678 and 687 (dating from the time of the first Gulf War) were insufficient in themselves. As time passed and war seemed inevitable, Goldsmith's advice became less unequivocal, but even with only days before war seemed likely to erupt, he had not given clear advice that war would be legal.

Concern about the legality of the seemingly inevitable war provoked serious concern in the country as a whole. Elizabeth Wilmshurst, deputy chief legal advisor at the Foreign and Commonwealth office, resigned over this very issue. In the Ministry of Defence, the possibility that British troops could find themselves committed to an illegal war was a very real worry – and one shared by many members of the armed forces. The Chief of Defence Staff (CDS), Admiral Sir (now Lord) Michael Boyce was forced to seek formal assurances about the potential war's legality from Lord Goldsmith,[10] who as late as Friday 7 March 2003 had not given a definitive opinion (the war began on 19 March). The CDS was faced with a real ethical dilemma. In a democracy like the UK, the military are subordinate to democratically elected politicians. CDS was faced with a choice, that of telling his prime minister, Tony Blair (who would not have been in the mood to hear this) that he was about to pull the plug on UK military involvement, with coalition forces already on the start line, or risk himself and those under him being dragged before the International Criminal Court in The Hague for participating in an illegal war.

In the event, we will never know how this particular situation would have resolved itself, as Goldsmith finally issued advice on 13 March 2003 saying that there was, on balance, a secure legal basis for military action.[11]

The Report of the Iraq Inquiry was, however, very critical about the way this decision was reached and concluded that 'The circumstances in which it was ultimately decided that there was a legal basis for UK participation were far from satisfactory'.[12] This is a rather British understatement. We can conclude from what the Report says that the way that this important element of the momentous decision to commit the UK to a war in Iraq was handled was a serious failing, lacking both transparency, and scrutiny.

I now want to return to Algeria and one aspect of its 'revolutionary war'. At the time of the war, Algeria was part of France, not a colony, but fully integrated into the French political and administrative apparatus. This war was terrifying and bloody, with deliberate and widespread use by the Algerian rebels, the FLN, of indiscriminate bombings of civilians in order to fuel fear and insecurity. Eventually, it became clear to the government that the police force was incapable of restoring order and that the situation was getting out of hand. It therefore authorised the army to regain control, using whatever means it considered necessary.

The French paratroops who were given the job, although without experience or training in this sort of work, approached the task with grim determination. But, as the French army became more and more involved in the conflict, and as its members became more and more exposed to almost daily horrors, so it began a descent into a *modus operandi* in which the use of torture became systematic, even routine. The forces involved had, in effect, ceased to behave as disciplined soldiers, bound by the Geneva Conventions and international law, and acted as if they were above the law. Initially, however, the tactics had some effect, and Algiers was briefly restored to peace and the FLN leadership decapitated, so victory seemed to have been won.

The respite was short-lived and violence did flare up again, but the most significant change was that news of the methods used by the army started to leak out and began to turn the French public against the army. As General Royal says in one of the most telling phrases in his book, 'The French no longer recognised themselves in their army' (Royal 2012: 139). The FLN also skilfully exploited these revelations to undermine the case for continued French rule in Algeria. In due course, Algeria gained independence, and the French army's image in the eyes of the French population suffered a catastrophic deterioration which lasted a generation.

The use of torture in the Algerian war also brings home a valuable lesson about the importance of ethical conduct in conflict. The French in Algeria were arguably defeated not by an enemy that was ruthless, brutal and unethical, but by its own unethical actions; by politicians implicitly giving the military an unethical task which the military in their turn accepted and undertook to perform. Moreover, in the equation between the FLN and the French government there was no equivalency in the public view. The French public did not expect anything better of the FLN, but they did expect better of their own government and army.

Indeed, one of the most incredible aspects of the way the French behaved in Algeria followed from the real possibility that, as a nation, they had been deeply scarred by the German occupation of France in the Second World War: the brutality of the Nazi occupiers had traumatised the French nation. Yet within little more than a decade, a part of their own army, at the prompting of and with the knowledge of their own government, was using the same strategy and tactics: the French public's shock on discovering this can only have been all the more profound because of what they had recently lived through themselves.

Michael Gross, in his book, *Moral Dilemmas of Modern War* expands on this same theme. If a government can define circumstances as 'exceptional', then it can be argued that it can use exceptional measures to deal with them. For example, Al-Qaeda's global network of terror and the threat it posed to the USA in particular was declared to be so exceptional that exceptional measures were deemed justified, such as the creation of an entirely new category of prisoner 'the unlawful combatant', to be held in a legally ill-defined facility – Guantanamo Bay, and subject to extraordinary rendition without judicial process. For Gross, there are two dangers here. The first is that governments will try to 'shoehorn new practices into international law' (Gross 2010: 234–235), and the second is that the exceptional evolves to become the new rule[13] and therefore no longer needs to be justified.

The effects of unethical and unacceptable practices go wider than the units and the individuals themselves. The use of torture by the French army in Algeria hastened the strategic defeat of France, in spite of the army's initial tactical successes. What happened at My Lai played a significant part in turning US public opinion against the Vietnam War. The massacre by Serbs at Srebrenica galvanised the West to finally put an end to ethnic cleansing in the Balkans. So, once they become public knowledge, unethical acts can serve to fatally undermine your own cause.

And if you know that, you can rely on the fact that your opponents also know it, and will do everything they can to exploit your unethical behaviour. In asymmetric war, the weaker side knows how important it is to dominate the battle for public opinion. They will use whatever means they can to show you in the worst possible light, so only a fool would give them the ammunition to do so. A single proven unethical act can not only be damaging in itself, but can create a climate in which public opinion is prepared to believe reports of any alleged instance of torture, deliberate targeting of non-combatants and similar acts.

There were many Islamist viewers and listeners only too prepared to believe the worst of the coalition in Afghanistan. But before we congratulate ourselves on being more sophisticated than they are, we perhaps need to remind ourselves how willing we were to believe the often-uncorroborated citizen journalism that emerged from Egypt and Libya's Arab Springs and initially emerged from Syria. In the midst of a violent conflict and in an era of 24-hour news and citizen journalism posted to YouTube, media organisations may not have the time, resources, opportunity (or even inclination. in some cases) to check out the facts. In asymmetric news terms, 'getting your retaliation in first' is often better than issuing a ponderous denial, after the facts have been carefully researched, but, wait too long, the damage to public opinion has been done.

ISIL and its affiliates and imitators have become highly sophisticated in the use of digital media to spread its propaganda and attract recruits. It parades its barbarism as a badge of honour, justified by a delusional belief in its own theocratic infallibility. But this barbarity, which would not be out of place in the Middle Ages, does not prevent it from using sophisticated modern technology to influence and spread its message. It also knows that Western media outlets will inevitably seize upon its atrocities and provide them with millions of dollars' worth of free publicity in the interests of 'impartiality'. But ISIL's most energetic publicity is aimed at potential recruits. Those who wish to look can find articles and videos which present participation alongside them in their 'Caliphate' as the ultimate religious duty. Many impressionable minds have been turned in this way.

The West does not really know how to deal with this. In the past, potential radicals were relatively easy to spot from their attendance at demonstrations or membership of fringe associations. Today, the potential recruit often sits alone with his (or her) computer in his bedroom, actively discouraged from making their true affiliations known and communicating using encrypted media. This has led to one of the great ethical debates of our generation, whether the potential interception of all electronic communications is justified in order to prevent atrocities committed by a small number of individuals.

Let me summarise at this point what I believe the main ethical risks for individuals in the era of wars amongst the people, before I turn briefly to what can be done to try to meet them.

The first challenge is that of war itself. Many peacetime norms have to be put aside and replaced by a maelstrom of tensions, difficulties and physical and mental struggles, all of which combine to challenge the ethical well-being of the combatant.

This leads to the second challenge: that the abnormal becomes the normal. Some 'abnormal' acts, such as the killing of another human, are legally sanctioned, but only if undertaken in accordance with the laws of armed conflict. Other abnormal acts, such as the deliberate targeting of non-combatants, remain firmly outside legal sanction.

Third, the temptation to go beyond these legally sanctioned acts can drive people to take the law literally into their own hands, particularly when the enemy has little or no regard for the Laws of Armed Conflict/International Humanitarian Law.

Fourth, this can lead to a risk that the enemy is no longer seen as a person, but 'demonised' as merely an object without feelings or inherent human dignity.

In Chapter 2 of his book, General Royal identifies three principles to guide ethical behaviour in response to these challenges. These principles are: fight hard, kill when you must, but respect life; unconditional respect for human dignity; cultivate the absolute primacy of example.

These principles seem to me to provide good and comprehensible guidelines to govern behaviour in the difficult environment we are talking about: they can be expanded upon as follows.

Fight hard, kill when you must, but respect life. On military operations, and especially in combat, only 100 per cent commitment will do. Being prepared to kill – or be killed – is an essential part of preparation, especially as most humans have a natural reluctance to kill their fellow human beings. But killing cannot be indiscriminate or unjustified.

Unconditional respect for human dignity. Note that this is not merely respect for human dignity, but *unconditional* respect. We saw in the Algerian example what can happen to an army when some humans are deemed to have less dignity than others. The same happened at My Lai and Abu Ghraib.

And finally, always set the right example. Commanders – and everyone above the rank of private is a commander – must lead by example. This is true of all aspects of the profession of arms, not just the ethical ones. A commander who is vengeful, proud, indifferent to the humanity of others, unwilling to treat the enemy with respect and has no self-control, cannot expect those who serve under him or her to behave any differently.

Leadership is hard to define,[14] but easy to recognise. But the way a leader is seen is a picture that is built up over time, based at least as much on how he behaves as on what he says. The foundations of ethical leadership are laid every day by everything a leader says and does. They are laid in peacetime, during training, by the respect shown to those in the chain of command above and below the leader. Ethical leadership begins by earning respect and is about inspiring those under command to follow, not making them fearful.

It will be the quality and nature of leadership that determines how those who serve under command meet the ethical challenges they will face. That is the responsibility of command and the first and most important ethical challenge that must be met.

So, in this turbulent ethical landscape, what can be done to prepare commanders and those under command to deal with complex ethical issues, knowing that they will arise? And preparation is essential; no commander would contemplate going into battle if physically unfit or without the knowledge to operate his weapon. It would clearly be totally unrealistic to begin to train troops on a new weapon system in the middle of battle; that training has to be undertaken well before combat begins.

However, for the weapons system, everything can be written down in a manual. Assembly, disassembly, what to do if the weapon jams, these actions can all be practised repeatedly until they can literally be accomplished blindfold.

Not so with ethics, where the groundwork, by way of ethics education, must be laid well before operations begin. Every situation is unique, so it is not possible to teach someone exactly what to think or what to do. They must be taught how to think through the complexities of an issue and the likely consequences of their actions. Such education takes time, as well as leadership.

But every training scenario can throw up an opportunity for education in ethics; such opportunities do not have to be created artificially. Those opportunities must be exploited. In military education terms, ethics is an unusual subject, because it can be studied both theoretically and separate from other, more practical subjects. But it can also be incorporated into most training scenarios, directly or indirectly. A huge amount of responsibility is placed on the shoulders of young men and women of junior rank who are placed in situations of great stress. Their training and education in ethics can only take place, like any other training, in controlled situations away from the pressure of the front line.

And such training has to be carried out in the knowledge of the pressures that will be faced in combat. So, training has to be thorough and committed. Simply going through the motions in a classroom will not be good enough. This is clear from the results of a 2006 report from the US Surgeon General's office,[15] which found (*inter alia*) that:

> Only 47 per cent of US soldiers and 38 per cent of US marines agreed that non-combatants should be treated with dignity and respect.
>
> Well over a third of the US soldiers and US marines felt that torture should be allowed, to save the life of a colleague or to gain important information about insurgents.
>
> Although they had received ethical training 28 per cent of US soldiers and 31 per cent of US marines reported facing ethical situations to which they did not know how to respond.
>
> Combat experience, especially losing a team member was related to an increase in ethical violations.

This suggests that however thorough the training that they had received, the US marines and soldiers surveyed were not prepared for the assault on their own ethical foundations which the horrors of combat, as described in the quotation at the start of this chapter, unleashed.

The key message of this chapter is therefore that everyone exposed to the nightmare realities and stresses of war will unquestionably face ethical challenges when on operations. The time to think about these challenges, to prepare for them and to set one's own moral compass, is now.

I would like to conclude by reminding you of the telling phrase that I quoted earlier from General Royal's book: 'The French no longer recognised themselves in their army' (Royal 2012: 139). For the citizens of a democracy not to recognise themselves in their own armed forces is a savage indictment indeed. The French army is now a superbly professional force, which takes ethical education very seriously. It does so in part because it understands the consequences of not doing so; of starting on that downward spiral that leads to it alienating itself from its own society and attracting opprobrium to itself. It has not looked into the abyss; it has been in the abyss and has had to redeem itself.

Those who serve in the armed forces of all countries, and the politicians to whom they answer, would do well to ponder these facts and to act on the lessons to be learned from them.

Notes

1 I use the term applied humanitarian intervention to describe deliberate armed intervention in the affairs of another state(s), as opposed to 'pure' humanitarian intervention, such as the apolitical provision of aid, usually at the request of the host government, following a natural disaster.
2 *The Baha Mousa Public Inquiry Report*, 2011, London: The Stationery Office.
3 For full text see Royal 2012, 40–41.
4 General Benoit Royal, 2011, *The Ethical Challenges of the Soldier – The French Experience*, John Thomas trans., Chapter 2, p. 41, Paris: Economica.
5 Ibid.
6 The situation was in large part legally resolved by an ROE change which allowed such suspected scouts to be engaged if they failed to drop their mobiles when challenged, but this did not wholly remove the ethical issue of potentially opening fire on children.
7 See House of Commons Defence Committee Report *UK Armed Forces Personnel and the Legal Framework for Future Operations Twelfth Report of Session 2013–14* published 2 April 2013.
8 *The Report of the Iraq Inquiry* (The Chilcot Report) Executive Summary para. 821.
9 For a fuller account of these events, see Michael Burleigh, 2011, *Moral Combat*, London: Harper Press, 304–307.
10 See John Kampfner, 2004, *Blair's Wars*, London: The Free Press, 304–305 and 378–379.
11 See the *Report of the Iraq Inquiry* (the 'Chilcot Report') Executive Summary para. 432ff.
12 Ibid.
13 See Gross, op. cit. p. 3.
14 Over 70,000 books on leadership have been published and there is still no agreed definition!
15 Surgeon General's Office, *Mental Health Advisory Team (MHAT) IV Operation Iraqi Freedom* 05–07 Final Report, 17 November 2006, 05–07 cited in Elizabeth Quintana, 2008, *The Ethics and Legal Implications of Military Unmanned Vehicles*, London: Royal United Services Institute, 12.

Bibliography

Burleigh, M. (2011) *Moral Combat*, London: Harper Press.
Coker, C. (2008) *Ethics and War in the 21st Century*, Abingdon and New York: Routledge.
Gross, M. (2010) *Moral Dilemmas of Moral War*, New York: Cambridge University Press.
Kampfner, J. (2004) *Blair's Wars*, London: The Free Press.

Quintana, E. (2008) *The Ethics and Legal Implications of Military Unmanned Vehicles*, London: Royal United Services Institute.

Royal, Gen. B. (2012) *The Ethical Challenges of the Soldier: The French Experience*, trans. John Thomas, Paris: Economica.

Smith, Gen. R. (2005) *The Utility of Force: The Art of War in the Modern World*, London: Allen Lane.

8 Challenges in combining ethical education for conscripts and professional military

The Finnish point of view

Janne Aalto

Introduction

The Finnish Defence Forces (FDF) is a military built on general (male) conscription. Annually about 75 per cent of men in their age group complete military service. This translates into approximately 23,000 people.[1] Whilst conscript service in the military is not mandatory for women, annually approximately 400 women still choose to complete the same training as men. There are no restrictions (such as a bar on front-line service as infantry, in armour etc.) on where women can serve[2] (Defence Command Finland 2014 2015).

Even though Finnish society has become more diverse, general conscription for Finnish men still enjoys widespread support, even among the young (Myllyniemi 2010). The support is significantly greater by comparison with other countries where conscription is in force (Rimpi 2007: 4). Finland is different from many other European countries in this regard, even though the recent developments in national security issues[3] have made many countries reconsider re-establishing conscription. General conscription in Finland has historical and geopolitical roots, and Finnish people regard it is a meaningful part of their national identity (Myllyniemi 2010). Therefore, the nature of the system in Finland and the aspirations of its citizens also affect the possibilities and goals of ethics teaching. Teaching ethics to conscripts not only involves teaching military ethics but also includes elements of citizenship education: it is hoped that such teaching will therefore affect people's way of life after, as well as during, their military service. Although contemporary weapon systems increasingly require the inculcation of more professional and specialised knowledge and skills (Finnish Defence Forces Deputy Chief of Staff, Personnel, MajGen Honkamaa 2012), the future capability of the FDF is still being developed against such a background of general conscription.

While the Finnish Defence Forces' wartime strength is approximately 230,000 soldiers, the peacetime personnel number of salaried personnel is approximately 13,000, out of which only 8,000 are actually soldiers. It can therefore be said that in times of a crisis the Finnish Defence Forces is wholly dependent on reserve personnel trained during their conscript service. Conscripts serve for between 165 days and 347 days, depending on the duties they are trained for. Naturally, the 40 per cent of conscripts that are selected for leadership duties serve the longest. After completion of the required conscript service a reservist can sometimes be ordered to return in later years to undergo 'refresher' exercises that can last from a few days to a few weeks.[4]

Of the small number of professional soldiers who serve in the FDF, approximately 3,000 are officers, 2,000 are warrant officers and 2,200 are noncommissioned officers. Of these, approximately 1,500 are officer specialists such as engineers, doctors or chaplains (Defence Command Finland 2014–2015). However, all professional soldiers, from sergeant to general, will have completed the normal conscript service. They will have started out as recruits, served 347 days as conscripts and in that time, been trained as reserve officers or noncommissioned officers (NCOs). After their national military service, the NCOs enter the reserve as corporals or sergeants in the Army and Air Force, or as petty officers in the case of the Navy: the reserve officers are commissioned as Second Lieutenants (or Ensigns) in the reserve. Only after a person has been given their commission as a reserve officer, or he or she has completed NCO training, can they apply to the Military Academy to become professional officers and receive the ethical education given to professional officers.

Those who do not become professional soldiers must rely on whatever (military) ethical education that they have received as conscripts, even though this is not much! And yet, on active service in times of crisis (and partially in refresher training) the professionals and reservists work side by side. Can we assume that they have a shared vision of the ethical principles that they follow in their activities? The goal of this chapter is to present how the FDFs have resolved this issue.

Follow the law and regulations – ethics education for conscripts

In this chapter I will, for the sake of clarity and that of comparison, look at the ethics education that a typical young Finnish conscript, who is trained as a reserve officer, receives during his or her service.

During his or her 12-month service, a conscript trained as a reserve officer will receive a total of six hours of ethics education. In addition, he or she is taught the laws of armed conflict and the code of conduct of the Finnish soldier, which is based on those laws (Defence Command, Training and Education Standards and Regulations). It is easy to see that you cannot perform miracles in such a short amount of time, especially when four of those hours are dedicated to teaching leadership and the ethics of leadership. Elements of citizenship education must also be included in these classes. In practice, this education, which is provided by the chaplains, is more or less forced to focus on the 'functional' level: we give and explain the regulations that must be followed and provide the conscripts with the boundaries they cannot cross. Of course, we try to encourage the conscripts to look at the issues from their own perspective, but they seldom do so. One cannot, therefore, exactly call this 'ethics education'.

An interesting detail in conscript training is the so called 'oath class'. Every Finnish conscript swears an oath or gives an affirmation.[5] The oath class explains the issues that the person swears to in the oath and what the oath obliges the person to do. It is worth noting that, although taking the oath is mandatory, it is in no way legally binding: abiding by the terms of the oath and upholding its standards are matters of honour and conscience rather than law. However, contemplating the oath during the 'oath class' often leads to ethics-related discussions amongst the cadets: the fundamental question for the oath class is: what are we committed to (as soldiers and as citizens)? Therefore, although a single lesson can only achieve so much, taking the oath and discussing its implications may well become a meaningful shared experience, as I will argue below.

Constandem decorat honor[6] – ethics education for professional officers

Just as with the conscript case above, for clarity's sake I will only use a single example of the potential career paths for professional officers and their education. A person can only apply to the National Defence University to study to become a professional military officer when they have received their commission as a reserve officer or completed NCO training. Those selected first study for three years for their Bachelor's degree in Military Science and, following this, they work in units as Lieutenants for about four years before returning to the university for two more years to complete their Master's degree. After receiving their postgraduate degree they are promoted to First Lieutenant and they may then advance in their career all the way to general officer rank.

During their studies, the cadets and Master's students receive ethics education. The scope, depending on their major subject, is 10–20 hours of classroom sessions during the Bachelor's studies and 6–30 hours during the Master's degree studies. In addition to this, they must read three–four articles about ethics and write a group essay based on those articles. There is no separate ethics session during field exercises but the instructors include the ethical aspect by asking 'what if?' questions. In addition, the students have the opportunity to write their final thesis on an ethics-related topic. Ethics education covers the basic theories of ethics (mostly the theories of normative ethics: deontological ethics, consequentialism, virtue ethics and the basics of value theories) and the applications for it in a soldier's duties, particularly within the frame of leadership and training and education. The laws of armed conflict and rules and regulations to be followed on operations are taught separately by lawyers in addition to the above. In terms of ethics, the goal of the teaching is 'aspirational': the education tries to stimulate the cadets' own reasoning and to give them the means to analyse their own identity as soldiers, and the means to consider and ponder upon the moral and ethical challenges that are a part of that identity rather than to necessarily provide definitive answers.

When a young man or woman has decided to become a professional military officer, he or she has to sit the entrance exams. The National Defence University's entrance exams also indirectly select people with similar values. The values are not officially researched or evaluated in the entrance exams, but the psychological tests and group discussions in the entrance exam phase eliminate the more extreme personalities. Overall, the applicant pool is small (approximately 600 every year) and exclusive as they have already been tested and evaluated for leadership training during conscript service. As students, cadets complete vocational and academic studies: both include studies on officer values and ethics. In addition to their studies, the cadets are also tied to these values and ethics by the officer community, which means learning the behaviour models (for example integrity, loyalty, politeness) and traditions of the officer corps. Of course, one of the challenges is to answer the question: is loyalty understood in the same way in 2015 as it was in 1935? Another aspect which is becoming more apparent is that the idea of the military officer profession being seen as a calling to serve your country is fading, and officership is now primarily seen more as a career and job, just as with any other career.

How, then, is the officer community seen? The ideal of a soldier can be looked at from many perspectives. It is often assumed that the ideal of the Finnish soldier is reflected in the military oath or affirmation (see Kaskeala 2008: 14, Leader's handbook 2012: 45).[7] Although the military oath can be seen as a kind of 'good citizen's' promise, you can also see ideals there that are generally considered a soldier's ideals or even virtues. These include reliability, fidelity and integrity. The military oath of allegiance places special requirements on those who are placed in a position of command. This is also practice that is common and accepted internationally: special obligations and ideals are often given and placed on officers. The Cadet Creed of the professional officer[8] highlights the 'officer virtues' of fidelity, courage, honour and comradeship.[9]

Naturally, the military oath and the Cadet Creed were not born in a vacuum, but have been affected by the historical, cultural and societal developments in Finland. The nineteenth century Romantic 'ideal of a soldier'[10] was fundamentally challenged by the new battlefields of the twentieth century, particularly in the 'Winter' and 'Continuation' Wars between Finland and the Soviet Union during 1939–45; wars that allowed much less room for the heroism of individuals (Jokinen 2006: 141–142). Despite this, some of those ideals have survived, even though the new situations and requirements of the wars have left their marks on the texts of the oath and the creed (Kallio 2009: 125). Latterly, there have been discussions on including a reference to duties above and beyond the defence of the homeland, such as peacekeeping and humanitarian intervention, in the Cadet Creed (see: Limnell 2009).

When reading the military oath and cadet creed it is easy to see echoes of a heroic society. Alasdair MacIntyre states that in an 'heroic' society, every position had its obligations and rights, and everyone knew which actions were therefore required and which were prohibited. When you evaluate an individual, you evaluate his actions (MacIntyre 1985: 149–150). In order to understand the virtues of a heroic society, you must understand that society's structure. In it, according to MacIntyre, morality and social structure are one and the same

(MacIntyre 2004: 151). A soldier must be brave, honourable, social etc., because otherwise he would not be a soldier. If he fulfils those expectations successfully, he rises to the position of a hero. He does not rise into that position because he followed ethical norms or acted to the benefit of others, but because he was successful in his role (Yrjönsuuri 1997, 106). The current virtues of a soldier still have echoes of such an heroic society, although in many other ways we live in a different era and the structures of society are completely different.

Aristotle's view of virtues is different from the picture painted in Homer's epics. For Aristotle, the virtues are not associated with individuals who fulfil a social role, but with the human (MacIntyre 2004: 217). Practising virtues is a way of reaching a goal, not an end in itself. Therefore, the fact that a soldier should be, for example, reliable, faithful and honest, has a reason. The virtues of a soldier are not the end in themselves but are rather based on qualities required to fulfil that role effectively.

According to David Fisher, the virtues of a soldier must be defined and it must be properly understood which parts of the virtues are needed or desired, because otherwise they can be misapplied (Fisher 2011: 123). Equally important is to define what a soldier's virtues mean in modern times. Fisher and Peter Olsthoorn have defined some of the old soldier virtues based on modern requirements (Fisher 2011; Olsthoorn 2011). However, it is not possible to do an all-encompassing generalisation of a soldier's virtues. Asa Kasher also reminds us that any conversation about military ethics must always be narrowed to some specific unit and its background, its meaning and position in that society and culture (Kasher 2008a: 134). That is why it would be good to further explain the content of the virtues in all those armed forces where a military oath is sworn or affirmed. If it is to be fair to the person taking the oath then it must be clear to him/her that they know what oath they are taking and what they are actually committing to.

In Finland, Häyry has defined the profile of a virtuous professional officer. Häyry begins defining the role based on the cadet creed and says that, following Aristotle, courage refers to the actual skill of being human: in other words, meeting the expectations of your role. Thus, according to Häyry, the starting point for a more detailed definition is meeting the expectations that people have of officers. He states that, traditionally, virtues have been seen as characteristics, tendencies to act in a certain way or as internalised principles. According to him, when interpreting them in this way, it is possible to teach virtues although, conversely, it is not possible to adopt them except through upbringing or teaching. He says that, in this case, if that is the aim of the exercise, creating a virtuous professional officer is indeed possible (Häyry 1991: 99–100). Still, the problem remains that in a military based on general conscription, not all of the officers are *professional* officers.

A test for the system: same duties, different background

In a crisis situation and also in crisis management duties, these two groups, reserve and professional military officers, operate side by side. What is needed for them to have the same approach to ethics at work when their backgrounds are so different? Let us take two young Lieutenants, Mikko and Jukka (these are common Finnish names). Both have completed a nine-year comprehensive school and a three-year upper secondary school and have received quite a similar education. After upper secondary school, they have entered military service, first as recruits and privates, before being selected for leadership training, during which they have been further selected for reserve officer training. For the last three months of their conscript service they have acted as instructors for the new recruits, under the supervision of professional military officers.

Although comprehensive school and basic training no longer guarantee that people have the same values, it does guarantee that these two individuals have common experiences on some level (Launonen 2000: 37). Also, the military service creates common and shared experiences and the leadership training given in the service also shapes the basis for any shared values. Hoikkala (2009: 154) argues that military service is a unifying experience for Finnish men. By doing his military service, a Finnish man symbolically takes his place as an initiated member of the male community (Hoikkala 2009: 155). Therefore, we can assume that when these young men go their separate ways, they take with them some shared experiences and a consciousness of shared values with the Defence Forces. Even though these values might be very 'everyday' and practical (Hoikkala 2009, 155), they still include a shared ethos that comes from such things as the shared oath.

Mikko has decided to become a professional officer and applies for the National Defence University. Jukka chooses a different path after military service. He ends up studying to be a primary school teacher (in reality, when I served in Afghanistan in 2010, of those reserve officers serving with me, nearly half were teachers. Others were lawyers, doctors, economists and police officers. One reserve Lieutenant was a carpenter by trade! Only the youngest Second Lieutenants in the reserve were without a pre-existing trade or profession).

No matter what field Jukka chooses, he will be taught the values and ethics of his own field and profession. If he goes to refresher exercises in addition to his studies or work, they will not directly teach him military ethics: rather, the intensive training is focused on his wartime duties (and, of course, the transfer and assimilation of military values happens there too).

After serving a few years as a Lieutenant, Mikko decides to request international duties and he soon receives orders to be deployed to the Lebanon. Since some duties are also open for reserve officers, and since Jukka (a Lieutenant in the reserve) also has an interest in this field he, too, applies for international duties and receives orders to go to the Lebanon.

In principle, these two men can thus occupy quite similar roles and undertake similar duties even though their educational backgrounds are completely different (I repeat that only professional military officers can be in the higher duty positions, but on the lieutenant–captain level the person in the combat uniform can be either a reservist or a professional officer). How, then, can we ensure that their ethical codes are the same and how can we train and educate them so that their 'operational models' are the same in situations requiring moral choices? To further understand what is involved here, I now turn to a consideration of the relevant pedagogy, via the notion of 'action competence'.

Soldiership and action competence

By the term 'action competence' I refer to an integrated concept and model of our mental, physical, social-cultural and ethical dimensions (see, for example Mäkinen 2011a). Although the concept of competence is strongly connected to the individual, it is not endogenous and formed purely by the individual, but is strongly connected to upbringing. Upbringing is always connected with the relationship between individual and community (Toiskallio 2009: 49). This connection affects the birth of an individual's identity, the idea of 'who I am'.

Action competence

Action competence (*toimintakyky* in Finnish) is just what it sounds like, the competence or ability to act. According to Toiskallio (2009: 48) this can be understood either so that the action competence precedes the action as a form of readiness, or so that it is the motivational force that is involved in actually carrying out the action rather than just thinking about it or being aware of what one should do. I will explore this further below.

'Action competence' should not be confused with performance. An individual's action competence cannot be measured or modelled in the same way as his/her performance. Still, in ever more complex military duties, action competence is the basis on which the capability of an individual, a section or that of a larger unit is built on. In a recent work, Toiskallio (2012) analyses this separation through the aforesaid English translation of the term. Toiskallio states that the word 'competence' is usually defined as an individual's ability to face the requirements set by the job – or, as you could also say, the task – by producing the expected result. Therefore, that person is competent for that job, task or performance. This, however, was not exactly what Toiskallio meant with his action competence model. *Competence* does indeed refer to the abilities and characteristics of an individual, but the result they produce clearly refers to *capability* (Toiskallio 2012). He then goes on to say that perhaps a better translation would be *action competency*, where the word *competency* refers to the individual's potential and special characteristics, such as knowledge level, abilities, motives, characteristics, sense of self and social role. Those special characteristics are key when reflecting on a soldier's ethical action competence.

Before reflecting on that, however, it is important to take a closer look at the modelling of action competence in general. A soldier's action competence, as military pedagogy understands it, means the entity formed by a person's physical, mental, social and ethical essence (Toiskallio 2009). The most important word in the previous sentence was *person* because action competence cannot be used to model a mechanical device or system, not even if it was equipped with the most advanced artificial intelligence, because you cannot consider them to be an independent personality that experiences things, has feelings or is morally responsible for its actions. Behind them is always a person, a human who is hopefully, for the most part, able to act and carry the moral responsibility.

The areas of the modelling can be separated from each other in order to observe them and to reflect on them, even though they can only function as a single entity. No part can be separated from that entity without harming the other parts or the whole. To explain it in a very simplified way: a soldier who is an extremely talented leader and has a total understanding of the social relationships of his/her unit and very precisely observes its (the unit's) atmosphere, but who cannot carry his own kit and equipment, cannot be said to have the necessary action competence. On the other hand, a soldier who is in top physical condition but is mentally volatile cannot be said to have the action competence either.

Next, to get to the heart of each, I will look at the different areas that make up action competence. The physical area of action competence is the easiest to understand, as are the possibilities for developing it. For example, if a soldier has to run 2,600 metres in a 12-minute running test and do 50 sit-ups in one minute and cannot do that, there are areas to improve in the action competence. Therefore, he/she should start training.

However, physical action competence alone is not enough. Mental action competence is connected to the physical in many ways; it is used to create that world of experiences that is partially built on our physicality. On the other hand, many mental functions, such as awareness of one's own alertness level, support the maintenance of physical action competence. Thus, mental action competence must not be understood 'only' as an element of perception or cognitive psychology. According to Kangas (2010: 112) it would be better to speak about *emotional* competence, because mental action competence also includes a person's existential reflection and understanding and the interpretation and experience of situations based on that reflection.

Still, a person does not observe, act, or do physical actions in a vacuum but in a world containing other people, communities and cultures. Mutanen (2010: 155) states metaphorically that 'an individual can only swim in water'. Social action competence is realised in the 'water', by identifying its operational models, modes of interaction, language and culture; it can be generalised by saying that social action competence deals with human relationships and interaction. Using military slang, it would be: 'Living in fully furnished accommodation'.

It is easy to understand that improving mental or social action competence – not to mention providing effective education on those subjects – is not as straightforward a process as improving physical action competence may be! Researching and educating people on these topics involves cooperation between many actors in many fields. According to Toiskallio (1998: 172–173) psychology and sociology are in a central position amongst the sciences involved, with additional contributions from behavioural science and sport and health sciences. Still, developing emotional and social action competence is a very long education process that is done in ongoing interaction between the individual and the surrounding community. At the same time though, maintaining emotional action competence can depend on very small things, such as communication possibilities with loved ones or good relations with members of the same unit, who are also a peer support group of sorts (Toivola 2011: 11–12).

Developing ethical action competence, the centrepiece of this chapter, is also a long process. With regards to action competence, ethics is not limited to 'clinical' ethics that can be taught in a completely neutral fashion. Being ethical, in terms of action competence, specifically requires 'action': ethical action competence is about *making* choices and decisions. It is active participation. According to Pihlström (2010: 172–173) ethics can never just be *taught*: it must also be *practised*. Pihlström's practice of ethics refers mainly to reflective ethics, but in ethical action competence the practising especially refers to making decisions about how I (or we) should act, so that my deeds will be just or good and so that I can accept responsibility for my actions (Toiskallio 2009: 64). In summary, it is about how ethical a soldier is – not simply that they know what the right thing to do is, but that they actually do it.

It is regrettable, though perhaps understandable, that in the training materials for conscripts the section on 'ethical action competence' is rather condensed: 'Ethical action competence is made up, for example, of awareness of morality: a sense of justice; a sense of responsibility: taking responsibility and being just' (Training material for conscripts 'Action Competence', Defence Command Finland 2014).

In the theory of action competence, the ethical action competence element, that is, how ethical the soldier is, is still a central part of the entire action competence model. It is only by making choices and decision that we make ourselves visible. Or, as Toiskallio puts it: we are realised in our ethical choices and commitments (Toiskallio 2009: 63). Hannah Arendt actually believes that describing another person is impossible, if we know nothing of their true actions: when we describe who someone is we describe what kind of person they are (Arendt 2002: 45). In the end, a soldier is also an individual, but if we start to define a soldier based on his/her reference group and our perceptions of it, the real identity of the person is lost. Still, a soldier's identity and how ethical he or she is strongly tied to each other because identity describes who is behind the decisions, what his or her values are and what he or she respects. Identity says who I am. Therefore, it is also important to look at a soldier's identity more closely.

Soldier?

James Griffith, who has studied the identities of reservists in the U.S. military, states that identity not only answers the question 'Who am I?' but also 'Who am I relative to others?' (Griffith 2011). This is of central importance when we examine the action competence model, because the core of action competence features the concept of identity, namely, in our case, the identity of soldiers or officers (Mäkinen 2011a: 81). So basically, it is about 'what it is like to be a soldier'. Related to the issue of soldiership is the question 'How can you retain your identity and ethics while you have to use violence and kill, as ordered by a democratic state that at the same time wants to uphold human dignity and human rights?' (Värri 2007: 32).

The same question of soldiership concerns both professional military officers and reserve officers, since in the end they are all members of the same military. In the beginning they have both been citizen-soldiers (see Mälkki 2010; Mäkinen 2011a). Although professional soldiers have a different kind of identity to that of reserve officers because of their long vocational training and professional education, they still need to place themselves in a military comprised of citizen-soldiers, since otherwise they will not gain the trust of their subordinates or their peers (Mälkki 2010: 244). One has to accept that a thought like this is possible only in a society that is still very equal and has hardly any class-consciousness. It will be interesting to see if future socio-economic changes in Finnish society will also be reflected inside the military, leading to an increasing division between personnel groups. Then we might perhaps truly speak of 'professional ethics' – a challenging concept when combined with general conscription!

Professional ethics

When discussing professional ethics, the concepts of *profession* and *professional* need to be clearly defined. A person that works in a job is not necessarily a professional. And the job that he/she works in, may not be his/her profession. According to Airaksinen a profession is different from a job, no matter how demanding and specialised the job. Still, Airaksinen continues, a professional is characterised by a wider set of skills and knowledge, without which he or she would not exist as a professional. This alone is not enough: a profession is also characterised by the point that it provides a way into a position in society. In addition, states Airaksinen, a real profession and professionalism also bring with them rights and power. Rights, in this case, refer to the right to organise the internal matters and conditions of the profession in a way the profession desires. Power refers to the autonomy a professional has in their own decision-making. Airaksinen combines professions, their rights and power with the concept of *authority*. Professionals are authorities in their field (Airaksinen 1991: 25–26).

However, authorities do not come out of nothing. One must ask why society gives certain employee groups such positions of authority that are effectively released from part of society's normal supervision and given greater freedoms to act as a group within a certain area of activity. Airaksinen says that the basic premise is that the professions serve those value-goals that are accepted by the majority of the society's members, such as health, safety and security, freedom, equality and welfare (Airaksinen 1991: 27).

Professions and their special positions can, therefore, be justified as long as they uphold the relevant values. Values, relating to who and what is supported and served, are generally seen as furthering some important societal goals. For example, the activities of doctor-support health, teacher-support education, officer-support security etc., although it must be noted that other employee groups also work for the same value goals, but may still fail to achieve such authority position. For example, nursing assistants, classroom assistants and the workers of security companies also promote the aforementioned values, but still do not get the same kind of respect in society as doctors, teachers or officers do. They may still have professional skills and some rights and power, but they do not have the same kind of freedom to organise their own matters as those groups that definitely constitute *professions*. By 'profession' I mean, in this case, a group in which expertise is based on academic education, and systematic and wide professional skills and knowledge (Juujärvi *et al.* 2007: 9).

Juujärvi *et al.* also propose that another identifying feature of a *profession* is a professional ethics code that defines the value-basis of the profession and the ethical principles of the work. Following from this is the fact that most, if not all, professions are self-regulating, in that breaches of the relevant codes of conduct are dealt with internally by the governing body of the profession in question. As with other professionals, the emphasis has been on having a strong identity and an ethically sound work ethic, and on being of service to the profession itself as well as to the section of the public that uses the services of such professions. Those elements have been suggested as sufficient to ensure that professionals always act in an ethical manner (Juujärvi *et al.* 2007).

However, changes in society have led to changes in the entire field of working life. Juujärvi *et al.* say that carrying on a trade is moving towards the holding of professional expertise. This means that the importance of knowledge-based skills is increasing. Everyone, not just a privileged elite, is expected to be able to reason through ethical problems. This situation can challenge the claim of traditional professional ethics to have a monopoly on ethical behaviour. This development may be a partial reason why there has been a perceived shift from talking solely about *officers'* professional ethics to talking about *soldiers'* professional ethics.

Soldiers' professional ethics

At this point it is necessary to consider briefly the general aim and objective of educating soldiers to behave in an ethical manner not only in wartime but in other situations involving the deployment of military forces (e.g. in peacekeeping and humanitarian intervention operations). This essentially involves adopting the guidance offered by the Just War Tradition, in which there has to be satisfaction of the moral (as well as the pragmatic) preconditions for 'going to war' in the first place (the requirements of 'jus ad bellum'), and then compliance with the conditions for fighting in a morally permissible manner for the duration of the conflict (the requirements of 'jus in bello'). Such is the only background against which discussion of the notion of 'military ethics' makes sense: indeed, Matti Häyry argues that without a clear link to such background, any discussion of soldiers' professional ethics is a waste of time. Given this explication, we can now turn to a more detailed examination of what is involved in creating a true professional identity for the soldier.

The Finnish Defence Forces Act defines professional soldiers as persons who serve in a military office/tenure (2007/551). The General Service Regulation lists military personnel to include the above mentioned as well as cadets, conscripts, reservists on active duty and those serving in military crisis management duties (General Service Regulation: 13). It is obvious that all of these members of the military personnel group cannot, for practical reasons or in principle, have a common, single, professional ethic. Worth noting is that this group includes 1) professional soldiers, 2) 20-year-old young men who are there just to do their 'duty' for 6 to 12 months and after that start their 'real' life and 3) reservists, whose latest contact with the Defence Forces is from over ten years ago and who now have a civilian job and the ethics associated with that civilian job. It does not change into a soldier's ethics during a week-long refresher exercise, because they most likely do not have a soldier's ethical code. Still, in international literature and research, military ethics is included in the 'family' of professional ethics. 'Firstly and most importantly, military ethics is a species of the genus "professional ethics"' (Cook and Syse 2010: 119).

Looking at the aforementioned, it is easy to understand that the discussion on a soldier's ethics and teaching ethics is centred mainly on the ethics of 'professionals'. Military ethics is seen as the business of professional soldiers. In the case of Finland, thinking like this presents challenges. International comparisons can be made mostly in regards to teaching ethics to professional officers in Finland. Also, Häyry states that a professional officer is the most natural object of research when researching military ethics in Finland (Häyry 1991: 87). Even with professional officers that system is not completely comparable, because in the officer education of many other countries the perspective of ethics education is teaching professional ethics to professional soldiers, who will also lead professional soldiers in their duties. Those being led have had other types of training on ethics.

Mälkki highlights the challenge for the professional officers working in a military that relies upon on general conscription. They must, for their part, be able to evaluate and face their subordinates' values and ethical ideas that they bring with them from civilian society and their own professional groups (Mälkki 2008: 243–244). It is still good to remember that ethics education in conscription-based militaries is not exclusively concerned with confronting challenges and demands. One of the purposes of conscription based on the Finnish system, is that it is intended to ensure that the military is never allowed to become separate from the values and standards of the rest of society – there can never be a 'them and us' situation because it is all 'us' (Mälkki 2010: 68). We can, therefore, speak of the *professional* ethics of *professional* officers (and in the future surely also that of the professional NCO's) but, in all honesty, we cannot speak of a soldier's professional ethics at large in a situation in which the current Finnish Defence Forces are based on a policy of general conscription. The issue must be looked at from a different perspective.

Identity and values

Asa Kasher (2008a: 133), speaking from an Israeli perspective, argues that a soldier's actions must reflect the values of the military and the state and that he or she must act and operate within the limits of the treaties that the state has committed to. Naturally, the state and the defence administration have their own values that they are committed to, military units have their own values and soldiers are trained by lawyers to abide by the international treaties that Finland has committed to. But when reservists are called up to service, we cannot assume that they will completely abandon the personal values they have adopted while growing up, nor can we think that they erase all the values that are somehow importantly related to their professional identity.

According to Anttila, it has been difficult for many who have served in crisis management missions to define if their idea of their own professional role has changed during or after the service or if it is unchanged. This applies to both professional soldiers as well as reservists. Even though many of them feel that serving in challenging missions and circumstances has given them a new perspective and a better idea of their own abilities, as well as an increased understanding of operating in different environments, they have not felt as though their core identities have changed (Anttila 2012: 87).

So, when we train our officers for challenging duties, we are not actually training soldiers but rather citizens and individuals. Ethical action competence is tied to the individual's identity since our ethical choices show who we really are (Toiskallio 2009: 64). Therefore, we cannot affect the lifelong growth and education only with periodic training or education, but we can adapt the education to fit the customs of the community that they are usually, but not exclusively, committed to. I say this to stress that, in the case of both professional and reserve officers and whilst they are deployed for crisis management or similar duties, they set aside their normal, day-to-day community values (I referred to this earlier by saying that professional officers also need to 'find their place' in the new organization).

When we Finns are preparing our soldiers to face the unexpected in a given cultural setting amongst a local population, the soldiers, in their respective units, have to grapple with the novel and unfamiliar within themselves. Therefore, prior to even trying to make sense of the unknown, unfamiliar and unexpected aspects created by the expeditionary operation's context, the soldiers have to have a crystal-clear understanding of their identity and values as soldiers (Mäkinen 2011b: 117).

Identity can be analysed on three interrelated levels; social, personal, and ego identity. Each of these identities is formed in part through interaction with the national context, meaning that the national histories, different geostrategic positions, as well as the national strategic preferences in terms of security policy have been shaping and are shaping the social and personal identities of soldiers (Mäkinen 2011b: 118). And on that social identity, which shapes the individual's identity, we base the ethical action competence on when we send our soldiers to crisis management duties. Within that identity are the values that we want our soldiers to follow. Adherence to those values is a very different matter from simply expecting someone to rigidly adhere to instructions, orders and laws. The values of professional and reserve officers are different (Mälkki 2008: 247), but the small tension between those values creates a unique but effective combination. They both need to take the other's values and ethical opinions into consideration in their actions. Thus, we reach the goal of ethical action competence. Acting according to it is different than behaviour directed or ordered from the outside. It is about having internalised the process so that one can take responsibility for oneself and of one's relationship towards others and the entire world.

So, what is: respecting other people's opinions? Reserve officers face the professionals' values and ideas of an officer's duties and have to take these into consideration (Pipping 2008: 201) to be able to co-operate with the professional military officers. In turn, the professional military officers have to take into consideration the values and attitudes of their colleagues who come from the civilian world. It must be noted that at the same time these values and attitudes reflect the values and attitudes of the society that has sent both groups there for these tasks. They are those basic values that the professional military officers cannot bypass or say that these values are secondary to their own values. The Finnish Defence Forces are an established part of society and therefore the values of society must be reflected in the operation of the Finnish Defence Forces (Puheloinen 2009).

In order to function, the system requires that there is ethical education and appropriate upbringing for the whole of society. Naturally, upbringing starts at home and continues all through one's life in different places, both in school and at work. That is why the concept of comprehensive security has become more current in discussions, along with the concept of human security (e.g. Heusala 2011: 97). Comprehensive security, in this broad sense, is very prominent, for example, in the Finnish *Strategy for Security in Society*, which defines both domestic and international threats that can touch upon the lives of Finnish people. The topics range from marginalisation in childhood to international crime and terrorism.

Ethical education should also permeate through to all the actors in society, not just the security ones since, as Mäkinen (2011b: 114) points out, the actions are not organised only through a single system (such as the military) but the operating systems are interlinked and a shared understanding can come to exist between them, for example on the concepts of good and evil and right and wrong. An understanding like this cannot be created by an order or brought in from the outside, but it requires that the values and their importance are evaluated on all operational levels. As previously mentioned, determining the importance of these values is not endogenous but they can be adopted through socialisation and upbringing. Both Värri and Ropo (2010: 135) are therefore worried about the lack of support for ethics education and the formation of a moral identity to support this, both among the security actors and in Finnish society at large.

Conclusion

This chapter is not arguing that the Finnish model for teaching military ethics is unique. However, the special requirements of a general conscription-based military must be taken into account when teaching military ethics to soldiers. In Finland's case, it is also accepted that there is little merit in offering the entire range of military ethics to everyone, as there has to be a difference between what is taught to the conscripts in the rank and file and to those who will be leaders. This is also true regarding aspects that are brought up in reserve refresher exercises if they are compared to training professional soldiers. It is not useful or even good to actively present all of the points of view or arguments to everyone because the time that is available does not make this possible, and in that short time it would be only a cursory glance at the complicated issues. In such cases, it is better to give clear orders and operating procedures (see, for example. Kasher 2008b; Bradley 2009: 292–293).

Maybe also because of the reasons mentioned before, it is notable that the field of military ethics in Finland is very wide and scattered at the moment. It covers everything from academic reasoning to studying the law of war in a behaviouristic manner, from a ten ECTS (European Credit Transfer and Accumulation System) intermediate studies module to a one hour 'do not do this' briefing, from the problematising of soldiership to very clear opinions of what makes a good soldier. Breadth is not necessarily a bad thing, but if the goal is to develop the ethical dimension of the action competence concept and see it as something more than simply an annoying section of the curriculum that no one is really familiar with, then some policies must be drafted. This leads to the question of who can make those policies arise – who owns ethics education?

Many different kinds of individuals and communities may claim ownership, but in my view, some limits can be drawn. When referring to a profession, let the professionals decide on internal policies. Here, I will cite, for example, the medical ethics discussions on abortion and euthanasia. Doctors and other healthcare personnel naturally participate extensively in the ethical discussions of these questions, but they have a special responsibility to also think of the ways, and the ethical aspects, of how procedures can be done in practice. Still, the decision of life and death is so fundamental that this discussion cannot be left solely to healthcare professionals as it is something that the whole of society must participate in.

Here I see a link with soldiers and military ethics. War and peace are things of such importance to society that discussing and reflecting on it cannot just be left to soldiers. The entire society must participate. However, the soldiers who implement the practical measures can and must reflect on the ethics of the methods used. This naturally covers professional officers and other professional soldiers, but who will speak for the conscripts and reservists?

The task of the Finnish Defence Forces in this process is to explain to every conscript the basics of the values that it expects a Finnish soldier to follow. Those in leadership training must also be taught the responsibilities of a leader, since their example has a very large influence on what kind of operating model, and what set of values, a unit adopts and practises.

Therefore, it is my conclusion that as we send our soldiers to crisis management duties or to other crisis situations we cannot, in a short time, educate and train them to act ethically. Nor in the current Finnish system do we need to. It must be remembered that there is no connection between stating high ethical standards and actually acting in accordance with those values (Olsthoorn 2011). The training and education must strive for action – actually doing the right thing, not simply knowing in theory what is 'the right thing to do' (Toiskallio 2004: 111).

The result is that we must equip our officers, who all have their own ethical opinions, to operate in a jungle of different values. The officers bring their own values and ethical opinions to the mission and it is the job of the Finnish Defence Forces to monitor that those values are inside the sphere of values that Finnish society expects the Finnish Defence Forces' actions to have, both in peacetime and during crises.

Notes

1 Military service must be completed between ages 18 to 30 years. The most common age for induction is 20 years. Non-military (civil) service is an alternative to military service.
2 Under current regulations it is not possible for women to apply to professional military duties or training without first completing military service.
3 I am writing this in summer 2015.
4 Some are never ordered to exercises, some almost annually.
5 I, (name), promise and affirm before the almighty and all-knowing God (in affirmation: by my honour and by my conscience) that I am a trustworthy and faithful citizen of the state of Finland. I want to serve my country honestly and, to my best ability, seek and pursue her edification and advantage. I want everywhere and in every situation, during peace and during war, defend the inviolability of my country, her legal system of government and the legal authority of the Republic. If I perceive or gain knowledge of activity to overthrow the legal authority or to subvert the system of government of the country, I want to report it to the authorities without delay. The unit to which I belong and my place in it I will not desert in any situation, but so long as I have strength in me, I will completely fulfil the task I have received. I promise to act honourably and with integrity, obey my superiors, comply with the laws and decrees and keep the service secrets trusted in me. I want to be forthright and helpful to my fellow servicemembers. Never will I due to kinship, friendship, envy, hatred or fear nor because of gifts or any other reason act contrary to my duty in service. If I am given a position of command, I want to be just to my subordinates, to take care of their well-being, acquire information on their wishes, to be their mentor and guide and, for my part, set them a good and encouraging example. All this I want to fulfil according to my honour and my conscience.

(Government Decree 1443/2007)

6 Constantem decorat honour 'Honour is the reward of the steadfast' Motto of the Military Academy (the Academy is part of the National Defence University).
7 For the oath, see n. 6, above.

8 During their first year of studies the cadets commit to the following creed: In front of the Cadet Flag, in front of the highest virtues of a soldier: fidelity and courage, honour and comradeship I promise to sacrifice my work and life for my country. During my time as a Cadet and after it may my emblem be Finland's freedom and the happiness of our people. Let the heroic memory of our fathers be the guiding light of my actions. May the incentives and goals of my thoughts and deeds always be elevated and noble. I will not shy away from toil, nor battle, suffering nor death to redeem this promise. May God help me to remain steadfast on this honourable path.

9 The concept of *virtue* is used here by Liikola, for example. (Liikola 2008: 2).

10 The epic poem 'Vänrikki Stoolin tarinat' (1848/1860) (*The Tales of Ensign Stål*), written by Finland's national poet J.L. Runeberg (1804–1877) is still linked with many of the ideals of the Finnish soldier.

References

Airaksinen, T. (1991) 'Philosophical Principles of Professional Ethics', in T. Airaksinen (ed.) *Ethics of Professions*, Helsinki: Yliopistopaino (in Finnish).

Anttila, U. (2012) *Enhancing Human Security Through Crisis Management*. (Diss.), Helsinki: National Defence University.

Arendt, H. (2002) *Vita Activa* (Finnish Translation). Tampere: Vastapaino.

Bradley, P.J. (2009) 'Why People Make the Wrong Choices – The Psychology of Ethical Failure', in Th. A. van Baarda, and D.E.M. Verweij (eds.) *The Moral Dimension of Asymmetrical Warfare*. Leiden: Martinus Nijhoff Publishers.

Cook, M.L. and Syse, H. (2010) 'What Should We Mean by Military Ethics?' *Journal of Military Ethics*, 9(2), 119–122.

Defence Forces Personnel 2014 (2015) Defence Command Finland, Personnel Division. Helsinki (in Finnish).

Fisher, D. (2011) *Morality and War: Can War Be Just in the Twenty-First Century?* Oxford: Oxford University Press.

General Service Regulation 2009 (2009) Defence Command Finland, Personnel Division. Helsinki (in Finnish).

Griffith, J. (2011) 'Reserve Identities: What Are They? And Do They Matter? An Empirical Examination', in *Armed Forces and Society*, 37(4), 619–635, originally published online 28 February 2011 at http://afs.sagepub.com/content/37/4/619.

Handbook for Leaders (2012) Defence Command Finland.

Häyry, M. (1991) 'Dying is for Youngsters – Or Is It? Some lines of professional military ethics', in T. Airaksinen (ed.) *Ethics of Professions*. Helsinki: Yliopistopaino (in Finnish).

Heusala, A.-L. (2011) 'Comprehensive and Human Security', in *Science and Weapon 2011*, Yearbook of Finnish Society of Military Science. Helsinki (in Finnish).

Hoikkala, T. (2009) 'Armed Forces as Institution', in T. Hoikkala, M. Salasuo, A. Ojajärvi, *The Known Soldiers*, Helsinki: Nuorisotutkimusverkosto (in Finnish).

Honkamaa, S. (2012) Lecture to the Participants of the Seminar 'Values' in National Defence University given by Deputy Chief of Staff, Major General Honkamaa.

Jokinen, J. (2006) 'The Myth in the Service of War', in T. Kinnunen and V. Kivimäki (eds.) *A Man in a War*, Jyväskylä: Minerva (in Finnish).

Juujärvi, S., Myyry, L. and Pesso, K. (2007) *Ethical Sensitivity in Professionalism*, Helsinki: Tammi (in Finnish).

Kallio, J. (2009) 'Unified Officers from Pan-European University', in M. Palokangas (ed.) *Honor is the Reward of the Steadfast*, Tampere: Apali.

Kangas, S. (1999) *Peacekeeper Congregation as the Construction of the Defence Forces Ecclesiastical Work in 1918–1999*. (Diss.), Helsinki: National Defence University (in Finnish).

Kangas, S. (2010) 'Pastoral Care in a Military Context', in J. Mäkinen, and J. Tuominen (eds.), *Military Pedagogical Reflections*, Helsinki: National Defence University (in Finnish).

Kasher, A. (2008a) 'Teaching and Training Military Ethics: An Israeli Experience', in P. Robinson, N. De Lee and D. Carrick (eds.), *Ethics Education in the Military*, Aldershot: Ashgate.

Kasher, A. (2008b) 'Military Ethics Between Code and Conduct', paper presented in the 9th International Conference on Military Pedagogy.

Kaskeala, J. (2008) *The Finnish Defence Forces in the world of globalization*, in H. Niskanen and S. Ahonen (eds.) 'It's a Question About Humans – Values in the World of Violence. Hämeenlinna, Kirjapaja (in Finnish).

Launonen, L. (2000) 'Ethical Thinking in Finnish School's Pedagogical Texts from 1860s to the 1990s' (Diss.), Jyväskylä: University of Jyväskylä (in Finnish).

Liikola, J.-P. (2008) 'Foreword', in T. Siren (ed.) *As an Officer in Hundred Years Old Finland*, Helsinki: National Defence University (in Finnish).

Limnell, J. (2009) 'Despite of Punish or Reward', in M. Palokangas (ed.) *Honor is the Reward of the Steadfast*, Tampere: Apali (in Finnish).

MacIntyre, A. (1985) *After Virtue*, London: Duckworth.

Mutanen, A. (2010) 'Values, Ethics and Action Competence', in J. Mäkinen and J. Tuominen, J. (eds.), *Military Pedagogical Reflections*, Helsinki: National Defence University (in Finnish).

Myllyniemi, S. (2010) *Youth Barometer*, Helsinki: Ministry of Education and Culture.

Mäkinen, J. (2011a) 'Military Pedagogical Comments on the Expeditionary Mindset – A Finnish Interpretation', in H. Furst and G. Kummel (eds.), *Core Values and the Expeditionary Mindset: Armed Forces in Metamorphosis*, Baden-Baden: Nomos.

Mäkinen, J. (2011b) 'Changeable Defence Forces – Unchangeable Soldiership', in *Science and Weapons 2011*, Yearbook of Finnish Society of Military Science: Helsinki (in Finnish).

Mäkinen, J. (2012) 'Conscription as Part of Society', in *Science and Weapons 2012*, Yearbook of Finnish Society of Military Science: Helsinki (in Finnish).

Mälkki, J. (2008) *Gentlemen, Lads and the Art of War: The Construction of Citizen Soldier and Professional Soldier Armies into the 'Miracle of the Winter War during the 1920s and 1930s)'* (Diss.), Helsinki: Hakapaino (in Finnish).

Mälkki, J. (2010) *Art of War for Leaders*, Helsinki: Suomen Mies (in Finnish).

Ollila, M.-R. (1997) *Beyond Moral*, Helsinki: WSOY (in Finnish).

Olsthoorn, P. (2011) *Military Ethics and Virtues – An Interdisciplinary Approach for the 21st Century*, London: Routledge.

Pihlström, S. (2010) *Problem of Life*, Tampere: Niin & Näin (in Finnish).

Pipping, K. (2008) *Infantry Company as a Society*, Helsinki: National Defence University.

Puheloinen, A. (2009) Lecture to the Students of the Higher Staff Course Given by Former (2009–2014) Chief of Defence, General Puheloinen.

Rimpi, K. (2007) 'Foreword', in *Social Needs for a Conscript System*, Helsinki: Ministry of Defence (in Finnish).

Salo, M. (2008) *The Chosen One – Characteristics of Small Group Leader*, Helsinki: National Defence University.

Toiskallio, J. (1998) 'Theory of Action Competence', in J. Toiskallio (ed.) *Action Competence in Military Pedagogy*, Helsinki: National Defence University (in Finnish).

Toiskallio, J. (2004) 'Action Competence Approach to the Transforming Soldiership', in J. Toiskallio (ed.) *Identity, Ethics and Soldiership*, Helsinki: National Defence College.

Toiskallio, J. (2009) 'Action Competence as a Concept of Military Pedagogy', in J. Toiskallio and J. Mäkinen, *Military Pedagogy*, Helsinki: National Defence University (in Finnish).

Toiskallio, J. (2012) 'Action Competence'. Unpublished paper presented in National Defence University (in Finnish, in author's possession).

Toivola, R. (2011) *Ways to Support Mental Readiness in Russian Armed Forces*, Lappeenranta: Army Academy (in Finnish).

Värri, V.-M. (2007) 'Some Problems of Ethics in Military Education: The Question of Ethics in Military Space', in J. Toiskallio (ed.) *Ethical Education in the Military*, Helsinki: National Defence University.

Värri, V.-M. and Ropo, E. (2010) 'How to Be an Officer and Gentleman?' in J. Mäkinen and J. Tuominen (eds.), *Military Pedagogical Reflections*, Helsinki: National Defence University (in Finnish).

Yrjönsuuri, M. (1997) *Good by Nature*, Helsinki: Kirjapaja (in Finnish).

9 Evaluating military ethics education

Common values, specific contexts

George R. Wilkes

A range of militaries have forged new approaches to military ethics education, and as part of this they have also explored a variety of forms of evaluation of their ethics education provision. One of the most obvious factors behind the pressure to evaluate ethics provision more rigorously has been the high risk of public ethics failures in respect of effective force projection in peacekeeping and counter-insurgency operations. These new risks place a demand upon military institutions to show that their educational reforms produce calculating military professionals leading military units which are fully prepared to act with the necessary professionalism.

Debate over the reasons for militaries to evaluate their ethics education with care is as old as the modern military academy (an exceptionally useful multinational source for this being Barnard (1872)). The reforms to military education in the eighteenth and nineteenth centuries created vibrant debates about the use of multiple means for evaluating the effectiveness of military vocational development: written exams, oral exams, evaluation by teachers able to capture the stature of an individual more fairly and exactly, and evaluation by a variety of examination boards – some internal to the military, some external – designed to ensure objectivity, competition and the rigorous application of high standards. New military challenges or crises returned the armed forces of European and North American states to the expectation with which military academies were forged: the formation of capable military professionals able to surmount the social and institutional limitations responsible for defeat in war. Militaries developed sophisticated examination systems geared to the promotion of quality officers and NCOs. The assumption that moral qualities could be inculcated through personal relationships with instructors, through corps discipline and through the teaching of other subjects meant that attention to moral qualities was an elusive feature of these examination reforms. Nevertheless, assumptions about ethics were also evident in many parts of the curriculum, and in attempts to make examination more rigorous and more effective as an instrument for overcoming social or institutional problems associated with poor quality amongst the students graduating from the academies (Barnard 1872).

The present contribution underlines how greatly evaluation techniques appropriate to assessing 'ethical preparedness' as a feature of overall force projection may differ from techniques more appropriate to assessing the development of ethical frameworks for the individual military professional.

The essay is divided into three parts. The first part examines the application of different evaluation tools: more subjective, personal evaluation by teachers personally familiar with their students to assist in vocational development, more collective or objective means, including external reviews, to address the high risks of massive ethical and public relations failures in the field. Those closely involved in evaluating ethics education are commonly involved in both, conscious of the different pressures involved in attempting to test vocational and situational 'preparedness'.

The second part extends this examination through a reflection on the different approaches to vocational and situational preparedness taken across military ethics and ethics education programmes. Whereas a comparable set of values can be shown to be deployed in military ethics teaching across many NATO militaries (see, e.g. Robinson *et al.* 2008: 5–7), the contexts for ethics classes differ within and between national militaries in a number of respects. This prompts divergent approaches to assessment of both the situational and vocational dimensions of ethics education.

A final section examines challenges encountered in the process of evaluating ethics education. Here, the paper examines arguments for pursuing different options for evaluation in the face of factors which make evaluation of ethical preparedness particularly difficult. The paper concludes with reflections on the interests, challenges and opportunities involved in focusing either on teacher or on external evaluations, or alternatively in combining the different examination resources.

Tools for evaluation: universal principles and standards, unique students, and meaningful feedback

Evaluation as a technical activity spans from the utilitarian and quantitative to the reflective and interactive. The former is an unavoidable feature of military education, reflecting the interests of the institutions which provide resources and which demand effective results from the military's perspective. These institutions place a 'top-down' pressure for evaluation and reform, and choices of evaluative frameworks naturally follow from this. The more reflective, interactive mode of evaluation lies at the conceptual heart of ethics education as a vocation. While it is far from ignored in recent moves to use ethics programming to bolster military effectiveness, this more subjective mode of evaluation demands different techniques which correspond to different notions of effectiveness. The time available for evaluation work in military academies is limited, but the pressure to deliver quality graduates who are adequately prepared has meant improving evaluation techniques has been a recurrent agenda item throughout the history of modern military education. The instructor looking through the material presented by educational reform commissions in the mid-nineteenth century will find much that is familiar (Barnard 1872).

High risks follow from ethical failures in military life, encouraging some militaries to expend increasing time and resources on innovative systems of ethics evaluation. These systematic approaches to evaluation of ethical preparedness across military units involve new evaluation resources which may be designed on a scale much larger than the set of evaluation opportunities or resources available to the staff dedicated to ethics provision in military education institutions. The design of these new systems varies greatly, with the choice of internal or external evaluators presenting one of the key variables that affect evaluation outcomes. In this section, we review methods used in evaluation, beginning with tools available to internal evaluators and moving to external evaluators. It may be seen that the choice of categorising evaluators in this fashion already reflects an awareness of the importance of the subjective dimension to evaluation processes. In the three parts of the essay that follow, it will be seen that the relationship constructed between internal and external evaluators is of critical interest.

The formative relationship on which contemporary military ethics education rests is that between educator and student. The educator models the ethical demands placed on members of the military, according to their place in the hierarchy and according to the situations they are likely to face. At most levels, this is likely to involve reflection on principles and cases. Classroom time may be supplemented by field trips or guest lectures. None of the details covered matter more than the relationship between educator – who students must find a credible representative of military norms that make actual sense in the field – and students, whose diversity and engagement with the subject educators will either fail to grasp or be able to work with. A more deliberately personal or more deliberately democratic teaching style may reflect a recognition of the importance of student diversity and student engagement with the subject matter and with the teacher. There is a further component that will arise in evaluation exercises. Whether or not the teaching style is deliberately focused on relationship building and participation, a further subjective component in course design rests on the ability of an educator or educators to present a programme which students find compelling enough to act upon when it becomes relevant.

A formative relationship demands an approach to evaluation of its own kind. Attempts to understand the quality or impact of this relationship through external evaluation after a class or after a course cannot wholly replace the subjective evaluations of those involved in the classroom. In practice, the most natural resource for regular evaluation is the instructor, whose time for evaluation is generally, and sometimes severely, constrained. Instructors may be able to accommodate time for individual exchanges with students in class or outside class, enabling a degree of directed discussion aimed at unearthing the subjective elements which affect student learning. Student evaluation is secured in many educational systems through reporting to third parties – classically achieved through feedback forms as courses near their end. In recent decades, more innovative approaches have begun to complement these opportunities, which are more time-consuming but also more integrated into curriculum design and more deliberately formed in relation to learning strategies. The course experience may be described through anonymised teacher and student journals. The anonymous journal format allows for fuller accounts of the reflective and subjective dimensions to ethics classes, requiring a deliberate choice and time commitment by educators where it is not established as the norm. It provides a fuller evidence base than feedback forms delivered at the end of a course. Oral exchanges, journals and other forms of reporting can be included in a formalised process of formative assessment of students, helping instructor and student to discuss progress and continuing challenges. Evidence of the performance of instructors may be achieved through third party review – an in-class peer review, or assessment by external evaluators. Each part of this assessment is time-consuming, and many institutions adopt only a part of each component in evaluating the effectiveness of educational provision.

In common with other classes in the academy or training context, ethics provision is commonly tested by a formal end-of-course examination, in some systems supplemented by mid-term exams or written papers. In some systems, examination is written, in some it is oral. These may be pass-fail tests of the retention of knowledge, delivered as formative exercises during an ethics course for NCOs, or at the conclusion of the course. Where exams test the ability to solve problems or to cast light on the dilemmas involved in difficult cases, institutions will commonly demand or reward right reasoning, which conforms with institutional expectations and justifies effective responses relevant to situations that may be faced in subsequent service. Classes at most levels will nevertheless also involve reflection on unclear situations, where moral capacity is measured not in terms of right answers, but in more imperfect terms. Exams can accommodate this, with the caveat that educators will recognise the difficulties presented in assessing moral capacity. First, there are subjective elements. Where ethics is mainstreamed into other courses, as part of an air power class, for instance, then the subjective dimensions to the ethics component may be judged quite differently by examiners who have technical, kinetic, or broader ethical specialisms. Second, candidates may anticipate 'right' responses, preventing examiners from accessing the subjective dimensions which an individual may view as their own real view, or as the unstated, politically incorrect view of the force to which they belong. Examinations conducted in the academy are not straightforwardly reliable indicators for subsequent reactions or conduct in the field, under fire, or in the heat of the moment.

Nevertheless, in military academies across the world, examinations are normally taken to be essential features in assessing candidate development. They provide quantitative and qualitative information that is readily available and naturally interpreted as evidence of the effectiveness of ethics provision.

Academy graduates will commonly encounter ethical dimensions to preparedness reviews, and in a growing number of militaries this ethical component is treated deliberately. Reviewers are generally external to a unit under review, though not necessarily, nor are they necessarily unfamiliar with the particular educational preparation, traditions, or spirit of the unit they are reviewing. Since reviews in preparation for specific missions and regular preparedness reviews (often annual) are designed specifically to measure a variety of aspects of military effectiveness, a problem-solving, utilitarian or consequentialist ethical framework is natural. Reviewers may expect right answers, and questions may focus on correct retention of doctrine or other essential information rather than on subjective factors. Nevertheless, there may also be scope for more personal elements to ethics reviews, both in testing individuals and in assessing the quality of relationships across a unit. This is of central importance in regular Canadian ethical preparedness reviews, coordinated by the Defence Ethics Programme, whose reports are online (see Canadian Armed Forces Defence Ethics Programme). Preparedness reviews can present an especially effective vehicle for embedding ethical expectations in military units during service, and not only in training. There is an obvious advantage in integrating the evaluation of ethical preparedness into the unit's mission focus, whether in reviews conducted on a regular basis or in pre-deployment scenarios, at which point evaluations can be targeted to risks and demands likely to be faced in the field.

The military expecting to face challenges to ethical preparedness may additionally establish feedback mechanisms operational throughout a unit's service, on base and in the field. Again, Canada has led the way in formalising a confidential feedback channel, able to produce evidence of ethical challenges and dilemmas which are viewed differently by commanding officers and members of a unit serving under them. The expectation that this system would prove to be supported by commanding officers and trusted by complainants has nevertheless been challenged by revelations of the breadth of unreported sexual harassment in the Canadian Armed Forces in 2013, for which a separate and independent reporting mechanism was then established. In other militaries, chaplains may informally perform a comparable role, where their position outside a unit hierarchy and their ability to guarantee confidentiality makes them an obvious resort for soldiers. As long as they are trusted, their role as ethical safety valves can increase further in units where they are also close to commanding officers. Whatever its form, a confidential channel of this nature promises information about potential and actual breaches with a distinctive mix of objective and subjective dimensions. It reaches into social realms which examinations broach obliquely at best. It may open up assumptions about the real or practical ethical norms that are not taught in the academy, and will reveal different forms of qualitative information about ethical issues associated with relationships within units, relationships with outsiders, and relationships with civilian family members. The channel through which information comes matters for parties interested in questions about the potential it holds for releasing subjective and selective information.

Evaluations by researchers – sociologists, psychologists, educational researchers – may complement these internal evaluation mechanisms in various respects. Again, there are many potential challenges in working with their findings. As external evaluators, researchers may be able to publicly identify the strengths of ethics education with a different form of credibility. They may also be able to produce meaningful comparisons with other militaries or other professional services. The first challenge for external evaluations by researchers is enabling effective access, in terms of time spent with research subjects, and in terms of the quality of researchers' interaction with serving members of the military. Researchers working within military institutions may overcome some of the barriers to mutual understanding, while external researchers offer fresh perspectives. In either case, the researcher–student relationship can produce different forms of qualitative and subjective evaluation than the purposive evaluations achieved within the military education and preparedness evaluation systems. Whereas educators and officers will have a natural interest in the progress of students towards military standards, researchers have a natural interest in exploring the impact of pre-military or extra-military life on subject's reception of ethics teaching.

The ability of researchers to gain the trust of members of the military is a crucial factor in their ability to capture attitudes to ethical norms and educational forms that may be intentionally hidden from instructors and officers, or not fed into the military's own evaluation systems. Two intriguing examples are to the point. An Israeli researcher spent her military service as a psychological support officer for soldiers undertaking a tour of duty in Gaza during their military service (Elizur and Yishay-Krien 2009). On the basis of their evidently frank conversations, she was later able to report not only on the nature and reported causes of infractions of ethical standards 'in the field', but also identified factors which singled out the two members of the unit who had refused to participate in unbecoming conduct: these two were older, having been to university, and they also appeared to be the only two members of the unit who received regular communications from their families. A senior Canadian forces researcher, Lisa Noonan, interviewed members of the Canadian Military Police, and was trusted with a range of comments which indicated that a significant proportion of interviewees believed that the real but necessarily hidden ethos of the service was far more socially conservative than the politically correct version which had to be given over in training and in statements by officers (Noonan 2006).

Internal evaluation systems produce effects on members of the military which external evaluators cannot. They can enable a data set to build up which enables a military to assess the appropriateness of ethics instruction and reinforcement after training ends, which can evaluate existing strengths and which can indicate areas for improvement. To the extent that external reviewers involve cost and risk, they may not be deliberately sought out in building up evidence for the quality of ethics education. In many countries, external reviews have nevertheless been an integral feature of military academy education since the nineteenth century (Barnard 1872). Militaries continue to have a natural interest in external evidence that they are performing well in this sphere, as in others, and particularly in contexts in which highly public ethics breaches have led to demand for publicly available evidence that problems have been addressed. The next section addresses these contexts and the responses they have produced in military ethics education programmes.

Contexts which demand customised evaluation

The urgency with which ethics education is formed and evaluated to respond to new developments is a source of strength as well as a reason for careful evaluation. New environments focus ethics provision on widely perceived realities, they bring new arguments about concrete challenges and consequences to bear, and they bring the force of institutions behind the drive to promote ethical behaviour and standards. The institutional, consequentialist and individual frameworks for thinking about ethical commitments work on different conceptual levels and in practice require a negotiation between purposive and qualitative programming and evaluation.

Ethics programmes are customised in all manner of ways in order to maximise their effectiveness and their utility. Curricula can at most give limited time for military ethics as a subject in its own right. Courses, stand-alone classes or events are therefore tailored to have an impact on course participants in proportion to the seriousness of the subject, and in proportion to the public expectation placed on militaries to display their respect for ethical norms. Visits to institutions focused on past genocides, for instance, are chosen by some military educators on both sides of the Atlantic for the clarity with which they impress upon members of the forces the consequences of a departure from universal moral norms and international law (for tailored courses partnered with the US Holocaust Museum, see USHMM in the bibliography). Ethics material and methods are introduced into doctrine classes and other courses in military academies, both as a deliberate attempt at mainstreaming the subject and also in lieu of dedicated ethics classes in academies in which the field is not yet accepted as a priority for officer cadets and other cohorts. The diffusion of case studies and of the teaching of frameworks for ethical reflection throughout the curriculum raises the profile of ethics within the curriculum. It affirms that a military aspires to become a thinking and a learning institution, with ethics seen as a rich source of intellectual challenge and as a source of motivation for individual initiative. At the same time, mainstreaming ethics makes it more difficult to evaluate. Each of these purposive steps can also present a complication for attempts to evaluate ethical learning in qualitative terms.

Giving ethics the weight it is due primarily in functional terms has consequences for any military which seeks to ascertain how effective or useful ethics education can be: it gives evaluators direction, while also creating pressures to depart from an ethics programme focused on resistance to purely functional calculation about values, behaviour and qualitative relationships. This is evident from the militaries which have the most elaborate ethics education programmes, such as the Canadian and Singaporean armed forces (e.g. Lew 2008). In both Canada and Singapore, the evaluation of the effectiveness of ethics programming has been driven by annual ethical preparedness reporting at commanding officer and unit level. In armed forces for whom dedicated military ethics programming is a newer demand, annual military preparedness checks are sometimes the primary opportunity for introducing an explicit ethics component into the fabric of military life. A full preparedness reporting mechanism can be a significant motor for change in the development of a culture of ethical reflection. The preparedness framework underlines the functionality of a sophisticated commitment to military ethics. By the same token, an effective annual ethical preparedness programme can prove to be a motor for a relatively narrow approach to what is conceived as being 'practical' about ethics in the military. A preparedness initiative classically rewards conformity rather than reflective engagement. The pressure to avoid the ultimate sanction of failure in a preparedness exercise can helpfully motivate attention to the prohibitions of international law. It is less helpful for the drive to promote an ethical awareness which applies to situations where those legal prohibitions are silent or inadequate. To extend preparedness monitoring to encompass ethical calculation means both evaluator and evaluated working against the grain of this core feature of the assessment exercise. It is by no means impossible, but it does require deliberate attention.

Asymmetric theatres and various forms of policing strategies in managing conflict and post-conflict situations have encouraged further demands for a more functional military ethics education programme across NATO (Carrick *et al.* 2013) and beyond it (Gross 2014). The result is the creation of tailored programming which may coexist in different ways with courses designed around the teaching of core values, philosophical ethical frameworks and case studies. The special demands of policing missions encourage the development of new ethics programming focused on calculating consequences, often delivered as part of intensive pre-deployment courses, often at battalion level and also for much smaller specialist circles, for medics or for chaplains, for instance. In the past decade, peace-keeping experience has encouraged some military academies (in Austria, Croatia, and Slovenia, for instance) to seek to combine ethics preparation with cultural awareness instruction, both as a reinforcement to teaching about the treatment of the local population and insurgents, and as a support to soldiers working in a multinational context, and whose cross-cultural experience may be minimal. Pre-deployment courses may be of critical importance to the co-ordination of teaching about the ethics-related components of the doctrine used in multinational missions, whether there is an agreed set of rules for engagement or the forces have to learn how their competing approaches to ethics relate to divergent rules of engagement. Such courses would need to be extensive in order to foster a renewed capacity for ethical reflection where this has not already been the subject of substantial work beforehand. Evaluating the delivery of pre-deployment ethics instruction against the purposes for which it is instituted is not a simple matter of listing and checking the acknowledgement of key ethical constraints or objectives in the field.

Professional military ethics educators are accustomed to course objectives designed to entrench a sense of ethical imperatives at a deeper level, framed by the values of the military. This is the case whether the context is a course for senior commanders or for new recruits undergoing basic training. This value- or identity-based approach to ethics teaching suggests a coherent set of learning outcomes which are easily appraised. There are, however, complications for the evaluator. Ethics courses at both of these levels commonly explore the potential for tension between an ethics based on the soldier's professional identity and the pressure to adopt a primarily functional approach to what is practical. A values-based approach, or an identity-based approach, to teaching (what is presented as) the fundamentals of ethical thinking and practice entails further problems for assessing what is in fact practical. The evaluator is confronted with a host of questions which may reveal the contrasting intellectual or ideological perspectives at stake in defining an ethical practice in the military. Answers to the question 'what is really practical here?' will first of all differ according to the identity or set of values or virtues that is chosen to represent the firm basis on which a soldier's ethics is formed. Across NATO, as the studies of the Military Ethics Education Network have shown, there is a common enough list of virtues associated with professional conduct (Robinson *et al.* 2008: 5–7), but there is also a wide divergence over the nature of the military's professional identity and values.

In many academies, the basis for the military's ethical commitments lies in national values, even though these may be the subject of heated dispute across the population. Canada's ethics programming is clearly designed to recreate the public image of the military as a force for good, while the Singaporean Armed Forces have deliberately set out to use ethics teaching to promote the virtues of a unified national identity, with an eye on the diversity of the country's population. In Singapore, and in Indonesia, an avowedly secular approach to military ethics is seen as unifying, and this may also be said of a number of Asian, African and European militaries.

By contrast, in many post-totalitarian militaries the relationship between secular and religious ethics is the subject of a highly politicised dispute. In some cases, the development of a core ethics component to military education may appear to have suffered as a result of this, if not also because of other obstacles to its inclusion in the curriculum. For some armed forces formed in situations of recent or ongoing internal conflict, the very premise of a coherent approach to professional military ethics lies not only in an assertion of values shared with civilians but also in the separation between the military and society. The promise of a non-political military ethic is not a cure-all for militaries seeking to circumvent domestic political division. It does not do away with the potential for controversy over what constitutes the distinctive practical duty of the soldier: witness the controversy over the warrior ethic which has been evident within the US military academies, entirely based on contrasting perspectives to effective military practice (Olsthoorn 2011 gives a glimpse into the debate, see, especially, p. 269).

Disputes over the nature of military ethics pit liberal approaches to the use of force against their critics, and universalist frameworks for justifying military action against more determinedly nation-centred or force protection-centred narratives. This is not an abstract or theoretical discussion. As is clear in a number of militaries, different sectors of the forces and the academies associated with them – the Marine Corps University in the US, for instance, and the Naval

Academy – understand their professional ethic in different ways, and this is easily understood in ideological terms, not simply in terms of different functions or traditions. Similarly, at ground level, battalions, regiments and their commanding officers understand their professional ethic in light of different traditions of service and different approaches to military objectives. These, too, can easily appear to conform to different ideological perspectives when military performance in Afghanistan, or Iraq, is the focus of a service-wide debate. Evaluating the delivery of values-based ethics programming is, as a result, not as simple as the decision to place values at the heart of army identity or education. Aware of the political or ideological stakes associated with ethical performance, course participants may conceal their personal resistance to new forms of military ethics programming, as we have seen was indicated in Noonan's survey work in the Canadian Military Police. A host of factors conspire to complicate the delivery of values-based ethics instruction, and in response evaluators must account for not simply the learning of core principles but also for a much more difficult process of relating these abstract notions to the practice of the military profession. The educator teaching core values to the ranks and the challenges of conveying the consequences of these values will quickly become clear.

Designing processes for effective, complex evaluations

The first consequence of these practical challenges is that decisions about evaluation procedures can be sensitive and complicated, the second is that the benefits of deliberate attention to evaluation can be overlooked.

In Singapore and Canada, elaborate forms of ongoing evaluation have been introduced with the aim of being able to assess the effective absorption of ethics education. Such elaborate structured evaluation mechanisms are not under consideration in many militaries, where the rationales given for new forms of evaluation have not penetrated educational institutions. Without institutional support, few educators will have time for formal approaches to evaluation, if by formal evaluation we mean the vagaries of feedback forms, peer review or other forms of reporting and course assessment. Instead, many militaries assess ethics education primarily by exam and as part of preparedness interviews with commanding officers. These various forms of assessment are also designed to help course providers to answer the question: has this course in ethics been effective? The question we will address now is: are these modes of evaluation really sufficient for the delivery of effective ethics instruction?

The Canadian armed forces have a much more developed evaluation process than most militaries because these traditional forms of assessment appear in a Canadian context to be of limited value in judging effectiveness in two respects. Both are influenced by Canada's experience of highly public ethics failures during peace operations. First, it is deemed not to be enough to ask an officer to report on their own ethical commitments because so much depends in a military unit on the officer's relationships with colleagues and subordinates. The ethical preparedness review is therefore designed to help to ensure that the military conforms to the norms expected of a democratic armed force, encompassing relationships built on accountability and to some degree transparency (Canadian Armed Forces Defence Ethics Programme). The Canadian ethical review allows an evaluation based on information about the context in which an officer works, with careful provision of confidential reporting of inappropriate behaviour. Effectiveness here is judged in relation to the prospect of serious ethical failures and also by the needs of a military unit.

Second, traditional examination procedures were evidently insufficient as tools for promoting ethics at an institutional level. Traditional examination, whether written or oral, encourages officers and officer cadets to report the answers which they know are expected, rather than to reflect on their own practices or their sense of moral development. The Canadian armed forces sought to put in place a more elaborate ethical preparedness system that was aware of the intense pressure within the forces to place greater reliance upon unit loyalties than on loyalty to norms for the protection of enemy combatants, civilians or individual members of the unit. At present, the resistance of members of the Canadian armed forces to use the new reporting systems is a matter of open public debate.

Evaluation of ethical preparedness in this collective sense, and of the realities which encourage departures from the values expressly identified with the armed forces, is not only an operation for instructors – nor only for the force's ethics teams and preparedness evaluators – it also lies in the realm of the social scientist or social psychologist. This requires the deliberate direction of research funds and manpower in support of the development of ethics education in the military: it is much less evident than the high-risk nature of ethics failures might suggest would be natural; but unsurprising in a climate in which military budgets are constrained, cut, or undergoing reconfiguration to pay for new technological developments. A small number of useful studies have been directed to attitudes to ethics provision in militaries in Europe and North America. We do not have many studies which give an account of the effectiveness of particular forms of ethics teaching. It may therefore be the case, for instance, as far as we know, that at present there may be many ethics teachers who assume the teaching of core values to be the most appropriate educating tool for new recruits, while others assume this to be too abstract to make an effective impact on students' behaviour. Nor have ethics studies been published which relate learning to subsequent field experience, nor to post-deployment well-being (though PTSD studies, notably in the USA, have made links that demand further attention). Finally, further studies on the impact of an individual's prior experience and education would be of enormous importance for the evaluation of the effectiveness of ethics programming in the military, and studies of this nature have yet to be published. Take new recruits as an example again. Most militaries take recruits from a range of backgrounds, but our knowledge about the ways in which recruits relate their training to their prior background is still underdeveloped. That academies would deliberately relate their examination procedures to questions about the qualities of academy recruits was a self-evident proposal for military education reformers throughout the nineteenth century. It is currently subject to constraints on the availability of researchers with a specialist interest in student performance – the specialism is supported in the Netherlands and Belgian academies, while in the UK and the US such studies depend more on direction from interested parties responsible for studies in particular academies.

In many academies, the type of personal assessments that might provide evidence for such studies are more complicated than is normally attempted in seeking to inculcate the military's core values. This may be justified on the assumption that it is more important to present all candidates with core values in the same way than to identify the particular attitudes of officer candidates from distinctive backgrounds. Where educators have an interest, by contrast, they will know that in an increasing number of school education systems recruits will have been exposed to a succession of forms of participatory evaluation, making this a relatively straightforward addition for ethics courses in the academies and in training for the lower ranks. The commitment to this approach to evaluation is an investment in a learning style focused on character development through personal engagement and on the capacity to build effective relationships in professional contexts. The format may be more structured than the teacher–student relationships that were deemed central to educational reform in the nineteenth century academy, but the significance of this relationship for effective evaluation rests on similar intuitions about the challenges to character development and to its recognition in an institutional setting.

Conclusions

Evaluations are deeply embedded in the bureaucracy with which military academies deliver high quality education. They are a core feature of the academy's historic attempts to identify their most able students. Examination systems have also been subject to repeated educational reforms aimed at reversing the social and institutional problems associated with military failure. This chapter has indicated that some of the more innovative evaluative frameworks used in the delivery of military ethics education, intended to serve this natural interest in fighting failure through educational reform, do provide a response to top-down processes of reform without necessarily engaging fully with changing educational realities and with developmental needs unique to individuals. Before rushing to assert the need for a joined-up approach to ethics education, it is worth recapping the distinct perspectives which create the scope for deploying evaluation tools in this way.

Ethics education as a spur to the development of long-term vocational commitments focuses teachers and students on individual decision-making resources, practices and styles. This is pedagogically subjective in the sense that the teacher is focused on developing ideas and commitments in the student as a person. The burden of the educational experience is on the teacher–student relationship: at a minimum, the teacher acts as a model whose integrity and credibility is important for the student's attitude to the proper application of ethics in their military service. A pedagogically subjective or person-centred approach is not necessarily ethically subjective, in the sense of conveying a relativist perspective in problem solving. Teaching methods and case studies may nevertheless be chosen for their effectiveness in developing appropriate student responses which highlight situations where there is no obvious or single right answer. To evaluate the effect of this vocational educational process on an individual implies attention to changes in personal attitude which are dependent on a web of social relationships which affect an individual's relationship to their military career. To evaluate the formative effect such education has on military performance also requires a subjective approach to evaluation: beginning in the classroom with the teacher–student relationship, and following student experience through from pre-military life to service and deployment after graduation. Written examinations and feedback forms capture only a part of the educational process at stake. This was already a feature of nineteenth-century educational reforms: a teacher's perspective on a student should speak more directly to the interests of their institutions where the quality of student vocational achievements is concerned. The nineteenth century reformers saw the need to balance teachers' perspectives with external evaluation, as a check on the subjective judgements of a teacher and their institution (which might extend to prejudice and nepotism in a military judged to be failing).

A military seeking to change its performance in the light of very public ethical failures has an interest in this vocational dimension, but also in quick changes at a collective level, measured in other ways. In this context, the individual's conformity to collective commitments becomes urgent. Challenges are created by situations and the consequences of public failure. An external, objective form of evaluation accessible to the military hierarchy outside the classroom meets an essential part of the needs faced by militaries seeking to limit the potential for damaging ethics failures. As was recognised in the Canadian reforms, an external systematic form of review matches an essential part of the need to influence unit performance where the ethical capacities or behaviour of individual commanding officers is in doubt. Feedback mechanisms also become, in this framework, focused on objections that can be made on objective criteria, not on subjective judgements. As the Canadian experience also underlines, external reviewers can face significant challenges in seeking to gain the trust and perspective they need to represent the ethical preparedness of units and of individual officers accurately. The drive for a system to guarantee ethical standards depends on subjective factors which imply an investment in subjective forms of evaluation, whether these are conducted by educators or by outsiders conducting interviews and observing the educational process.

The present examination of problems and agendas for reform suggests a combination of evaluation methods and perspectives. Uncontroversial, in practice institutional pressures and resource stretch may mean it continues to be underdeveloped in most military education systems, and even in those making the largest investment in ethics reform. Combining methods requires negotiations, communications across insider–outsider boundaries, an awareness of subjective backgrounds, interests and sources of trust issues, and a broader sense of the ways in which external forces and evaluators are also implicated in an evaluation exercise. No single simple exercise will lead to an evaluation that captures all the challenges at stake, let alone that leads to a problem–proof ethics education system.

That notwithstanding, in positive, practical terms, this chapter presents an argument for the practical benefits of a renewed conversation based on evaluations from teachers, students, outsiders within the military, civilians with a brief for reform and external researchers. Perspectives could be of great practical benefit to the military today. The relationship between internal and external evaluators is an important one to take seriously. This can be described more clearly in terms of the limitations of each form of evaluation. The academic or outsider perspective can pick up much which internal reporting does not. Equally, much is lost where studies are constructed without a full immersion in the structures of the military and in the specific conditions of service in which individuals apply what they have learnt in the classroom. Evaluation in its more elaborate forms can be important in addressing the ambitious and often subtle expectations we place on ethics education in the military. Character development is a central objective of most ethics programmes, even where its nature is not agreed. Military educators know it is difficult to evaluate, particularly via any objectifying method. At the same time, we expect of ethics provision that it be sensitive to different approaches to practical decision-making in the contexts in which this is needed in the military. Here, a systematic response corresponding to the public challenges faced by militaries may increasingly be found in forms of evaluation in which members of the military engage in a participatory fashion, in a conversation and not through a simplified objective assessment alone.

In the overstretched or financially straightened situation faced by many military academies, it is natural that time-intensive evaluations fall off the agendas of educators and their institutions. But precisely at a time where priorities are under review and ethics commitments require argument and reformulation, evaluations can prove one of the educator's greatest resources. An integrated approach to evaluation could prove to be one of the most significant motors behind the promotion of an effective, integrated discourse about professional ethics in military affairs beyond the academy.

Bibliography

Barnard, H. (1872) *Military Schools and Courses of Instruction in the Science and Art of War, in France, Prussia, Austria, Russia, Sweden, Switzerland, Sardinia, England, and the United States. Drawn From Recent Official Reports and Documents.* Revised edition, New York, N.Y.: E. Steiger.

Canadian Armed Forces Defence Ethics Programme survey reports, online at www.forces.gc.ca/en/about/defence-ethics.page.

Carrick, D., Connelly, J. and Robinson, P. (eds.). (2013) *Ethics Education for Irregular Warfare*, Aldershot: Ashgate.

Elizur, Y. and Yishay-Krien, N. (March 2009) 'Participation in Atrocities Among Israeli Soldiers During the First Intifada: A Qualitative Analysis', *Journal of Peace Research*, 46(2), 251–267.

Gross, M. (2014) *The Ethics of Insurgency: A Critical Guide to Just Guerrilla Warfare*, Cambridge: Cambridge University Press.

B.C. Lew, Psalm. (2008) 'Preparing Values-Based Commanders for the Third Generation Singapore Armed Forces', in Jeff Stouffer and Stef Seiler (eds.), *Military Ethics*, Kingston Ontario: Canadian Armed Forces.

Noonan, L. (n.d.) 'Right Wing Authoritarianism and Social Dominance in the Military Police', Presentation to the IMTA 2006 symposium, online at www.internationalmta.org/Documents/2006/2006057T.pdfý.

Olsthoorn, P. (2011) 'Loyalty and Professionalisation in the Military', in Paolo Tripodi and Jessica Wolfendale (eds.), *New Wars and New Soldiers: Military Ethics in the Contemporary World*, Farnham: Ashgate, 257–271.

Robinson, P., de Lee, N. and Carrick, D. (eds.), (2008) *Ethics Education in the Military*, Aldershot: Ashgate.

USHMM (n.d.) online at www.ushmm.org/professionals-and-student-leaders/military-professionals.

10 Challenges to the professional military ethics education landscape

David Whetham

Professional ethical competence requires translating ethical principles from theoretical knowledge to practical action in the real world, and developing this competence is the role of Professional Military Ethics Education (PMEE). However, the ad hoc development of international PMEE poses some core questions that the volumes in Ashgate Publishing's Military and Defence Ethics (MDE) Series have attempted to address. What actually *is* the professional military ethic, do different countries agree on what 'good' military ethics looks like, and do ethical concepts translate between cultures? How is ethical competence taught, how is it measured and why does this differ from place to place? This chapter cannot answer all of these questions. There is on-going work in many different countries aimed at attempting to do this – this new focus upon military ethics education is, in part at least, a tribute to the success of the original Military Ethics Education Network in highlighting these issues and raising them on the international agenda.[1] What this chapter will seek to do is outline some of the current challenges being faced in PMEE delivery, focusing on the experience of the UK. One can see distinct areas of challenge; the first relates to what should be taught in the sense of the changing contemporary operating environment and the ethical issues that it faces combatants with, while the second relates to the question 'to whom military ethics should actually be taught?' and to the best way of doing this.

Why does ethics matter for the military?

The military profession, as with all professions, is defined and governed in large part by its ethic: the rules and behaviours by which its members conduct themselves.[2] Ethical failures by the military can have terrible consequences for a huge range of people, including the local civilian populations and the soldiers on both sides of a conflict, for the internal health of the military organisation itself, for the relationship between the military and society, and for the strategic utility of forces engaged on behalf of their political community.

The type of 'discretionary conflict' that currently appears to be the norm, where vital national interests are not obviously at stake, can pose different ethical and legal challenges for democracies when compared to wars of national survival, where the issues appear more black and white (as far as this is ever possible in war). The potential range of issues that must be addressed is also widening due to the varied types of activity the military can become involved with.[3] Peacekeeping or peace enforcement and humanitarian relief operations pose very different types of challenges to those found in 'traditional' high-intensity, state-on-state warfare. As one of the preceding volumes in the MDE Series attests (Carrick *et al.* 2009) counter-insurgency and irregular wars introduce a whole raft of ethical and legal dilemmas that need to be explored and resolved if campaigns are to be conducted appropriately and ultimately judged as successful.

One of the questions that often arises is: why is this an ethics issue rather than simply a legal one? Every professional military around the world is supposed to fulfil its international obligations and ensure that the Law of Armed Conflict is taught and refreshed each year to all serving military personnel, both in peacetime and in wartime.[4] If this is done properly, doesn't this make military ethics education redundant? Clearly there is a great deal of overlap between the two areas of ethics and law. However, sometimes the contemporary operating environment presents situations in which it might not be clear what the legal position actually is:

> [s]mall wars demand the highest type of leadership directed by intelligence, resourcefulness, and ingenuity. Small wars are conceived in uncertainty, are conducted often with precarious responsibility and doubtful authority, under indeterminate orders lacking specific instructions.
>
> (Department of the Navy 1990: 1–6)

Even where the framework of the law is absolutely clear, and the range of legally permissible options can be clearly identified, the law 'can rarely provide the actual answers' (Whetham 2010: 2). A decision still needs to be made as to which course of action one *should* take. Military decision-making requires the ability to answer questions such as 'Would such an action be lawful in this situation?', but it also needs people who will also ask 'This course of action is legal but is it actually the *right* thing to do?'. The best decision-making will therefore be informed by both ethical and legal considerations if the appropriate and most desirable outcome is to be achieved.

The type of issues that come up in military ethics sometimes do not have straightforward answers, thus the need for education rather than training in this important area. It is precisely the questions for which there are not black and white responses that need to be engaged with, thought about and discussed by the people for whom they are most pertinent – military practitioners.

Challenges for military ethics – the changing environment

The range of contemporary challenges to traditional ways of thinking about military ethics is broad and getting broader, not just for the UK but for all military forces around the world. While this chapter is not intended as a comprehensive survey of these challenges, it is impossible to discuss military ethics education without acknowledging that the subject is dynamic. It is perhaps useful to employ the very familiar criteria of the Just War Tradition to briefly touch on some of these different challenges to show how every area within it needs to be considered and interpreted in light of the changes in the contemporary operating environment.

Just Cause

This the usual starting point when looking at the Just War criteria, and the clearest example of this being satisfied is when a state finds itself attacked. The legitimacy of acting in self-defence is enshrined in Article 51of the UN Charter, which affirms that 'Nothing in the present Charter shall impair the inherent right of individual or collective self-defence if an armed attack occurs'. Coming to the aid of allies or the weak and vulnerable might also be included under this heading (thus the emerging idea of the international community's 'responsibility to protect' when a state proves itself to be unwilling or unable to carry out its core responsibility to look after its own population).[5] But how is one to respond to a threat that has yet to manifest itself? US National Security Advisor, Condoleezza Rice, in 2002, suggested that traditional notions of self-defence were no longer adequate for the new security environment of uninhibited actors with potential access to weapons of mass destruction (WMDs).[6] Clearly one cannot wait until the appearance of the mushroom cloud to act in such circumstances and yet, morally, the problem with pre-empting a threat that has yet to actually materialize is that it can all too easily turn the defender into the attacker. Getting the balance right (and providing appropriate advice to political leaders so that they can do this) is essential if one wants to be considered a legitimate actor. Perhaps more problematically, the emerging threats to the state do not lend themselves easily to traditional language or concepts – is a denial of service attack aimed at a country's stock exchange that wipes hundreds of millions temporarily from stock prices an actual attack or not? More specifically, if no one was actually hurt, is this sufficient as a Just Cause for determining a potential military response? If not now, what is the threshold – must one watch the crippling of one's economy without responding with the military tools of state defence as long as there are no directly attributable deaths? Does a Just Cause require a physical threat?

Right intention

Most people would accept that the motivation for an act has an affect on whether it is considered morally good or bad, so how is the concept of 'right intention' changing?

'Creating, restoring or keeping a just peace, righting wrongs and protecting the innocent would all clearly qualify as right intentions, while seeking to expand one's territory, enslave or convert others to one's religion, hatred or revenge would not.' (Whetham 2010: 77–78). However, states rarely do anything inspired by a single motivation – they are enormously complex actors with a huge range of interests. How many oil contracts or reconstruction deals invalidate a humanitarian operation? Turning it around, do the parents who have seen their children saved by an international rescue force really care if one of the intervener's considerations was long-term trade development? Professional soldiers in an age of 'discretionary conflicts' rather than 'wars of national survival' may find it difficult to square what they are being asked to do with the official reasoning presented to them by their governments. Governments can be coy about referencing geostrategic interests or the importance of keeping alliances healthy when seeking to justify their actions, and the disconnect between the officially stated mission and the resources supplied to do it can often be stark. However, while accepting that altruistic motives can always be questioned, this does not diminish the importance of recognising that 'wars fought *primarily* for the wrong motives will invariably lead to an unjust peace' (Whetham 2010: 77–78).

Legitimate authority

Determining what constitutes a 'legitimate authority' in the twenty-first century poses as many problems as it did in the Middle Ages (see Whetham 2009). On the face of it, this should be straightforward – unless acting in self-defence (which requires no further authority), a state requires the prior approval of the United Nations for military action towards another actor. In part, this explains why states are often keen to characterise the defensive nature of their activities. However, it is also clear that in practice, states claim legitimacy when acting through regional alliances. Are states the only legitimate actors in the international system? Does a government recognised by the international community but rejected by 98 per cent of the population it is supposed to represent, have any real legitimacy? Without any popular support, can a government really declare war? (Holliday 2002: 567). What about rebel movements – are they automatically illegitimate until they win, or is it just until they have the support of at least a majority of the people? At the point they have a degree of popular support, who becomes the real insurgent? From a PMEE point of view, how does a professional soldier know they are on the right side of a dispute, especially if that conflict is between different elements within the state itself, or is opposed by a significant proportion of the people that they are supposed to be serving?[7]

Proportionality

What does proportionality (at the *ad bellum* or the policy level) mean in the types of conflict that we are seeing this century? This prudential criterion demands that we ensure that the overall harm likely to be caused by the war is less than that 'caused by the wrong that is being righted' (Bellamy 2006: 123).

This has never been an easy calculation to make, but, returning to an earlier point, how is one to weigh the damage or harm caused by a cyber-attack? What actually is a proportional response to such a threat – political censure, diplomatic pressure, applying a law enforcement paradigm or moving into and along the spectrum of military responses?

Reasonable prospect of success

This is also considered a prudential criterion, one is supposed to ensure that there is a reasonable prospect of success before embarking on a military course of action. What does success in war look like in the twenty-first century? Does it require a victory, at least in the Clausewitzian sense of inflicting defeat by 'destruction of the enemy forces, whether by death, injury or any other means – either completely or enough to make him stop fighting' (Clausewitz 1989, Bk 8, Ch 2: 579)? If not, at what point is one supposed to determine success? Arguably, many of the problems that the international coalition encountered in Afghanistan were caused by a failure to determine what exactly they were supposed to be doing there. Was it about bringing to account those responsible for the crimes of 9/11 and preventing the country being used as a base for further attacks, neutralising a threat to the region, introducing democracy, supporting women's rights, or all of these things wrapped up in the all-encompassing phrase 'nation building'? The link between PMEE and effective strategy is nowhere clearer than here – Clausewitz made clear that: 'no-one starts a war – or rather, no-one in his senses ought to do so – without being clear in his mind what he intends to achieve by that war and how he intends to conduct it' (Clausewitz 1989: Bk 4, Ch 3: 227). Professional soldiers need to understand and ask the right questions from their political masters about what exactly it is that they are being asked to do. It is they, after-all, who will be held accountable when the judgement is made as to whether the blood and treasure was worth it.

Last resort

Finally, at the *ad bellum* level at least, the use of military means should be a last resort, but how is one to interpret that in the context that Condoleezza Rice sets out above? Waiting to satisfy the imminence criterion of the last resort would be suicidal in the face of some types of threat. How about inactivity in the face of an emerging genocide? Does one have to wait until the genocide has happened before one can prove that military action is the only viable option left that might work? If one acts to prevent the genocide and is successful, then how does one prove that the 'last resort' criterion was ever actually satisfied?

Discrimination

At the *in bello* level, professional military personnel must know how to apply discrimination in their actions. Understanding how to apply discrimination (or 'distinction' as the legal term is framed) to separate those who may be targeted from those who may not, even in a traditional battle space, is difficult enough, but what about when an enemy deliberately attempts to 'draw the foul' and create a situation in which whatever the other party does, it is perceived as illegitimate (Whetham 2007: 721–33). Examples could be using women and children non-combatants as human shields or deliberately employing tactics that undermine the rules of war, such as feigning surrender. If you are the party placed in such an impossible situation by an unscrupulous enemy, what can and should you do?

Proportionality is closely related do this principle. Again, the idea is simple enough: the total damage, losses and injury resulting from any military action, not just to one's own side but considered overall, must not be excessive when compared with the expected military advantage of the action. How is one to interpret Proportionality in an environment where international opinion is accustomed to precision munitions? How do new technologies such as drones change the moral landscape? What does proportionality even mean in a cyber environment?

These examples touch upon just some of the ways that the ethical landscape of warfare has changed. The issues raised above are just a small cross section of the ones that philosophers, ethicists, lawyers, policymakers and soldiers are grappling with right now. For example, there are debates as to whether the Just War Tradition is still even the most appropriate language for having these debates, or whether the principles need to be reinterpreted in light of changing circumstances. For professional military officers, engaging with these issues and understanding these debates is an essential part of the *jus ante bellum*, the area concerned with the moral, ethical and legal preparation that should be conducted *before* conflict is considered or entered into.[8]

To whom should military ethics be taught?

Chapters in earlier volumes in the MDE series have looked at the way PMEE is delivered at the different levels of training and education provided for the British Armed Forces (as well as for the armed forces of other Western democracies). Just as the ethical landscape of warfare is changing, so too is the landscape within which military ethics will need to be taught in the UK. As the above brief survey has highlighted, uncertainty is an inherent part of the contemporary operating environment. As a result, our service personnel are expected to make decisions, often profoundly important ones, without all of the facts being available to them. Clearly, trying to indulge in the luxury of a Rawlsian 'reflective equilibrium' while taking incoming fire, or trying to decide whether a search and rescue mission should be ordered when personnel are in the water but the flying conditions are atrocious, would be to test even the most experienced and skilled philosopher practitioner to the limit. Therefore, just as with any other kind of professional activity, it is essential to train in advance for the type of scenarios one is likely to be faced with. Dave Grossman puts it like this:

> You do not rise to the occasion in combat, you sink to the level of your training. Do not expect the combat fairy to come bonk you with the combat wand and suddenly make you capable of doing things that you never rehearsed before. It will not happen.
>
> (Grossman 2007: 77)

This way, the type of factors that need to be considered will already be accessible to the individual making the decision. While, obviously, every situation will be unique, it is also the case that many factors and considerations will be pertinent in a number of different scenarios. The ability to assess the context and draw on the right factors and weigh them appropriately is part of the ability to analyse – a skill that is at the heart of what Staff Officer education in the UK is about.[9] This helps to equip those officers with the necessary skills for moving into senior positions in the operational level of war – 'the level at which campaigns and major military operations are planned and sustained' (Whetham 2010: 1).

While the dream of network centric warfare was to remove the fog of war and ensure that the commander was able to direct and control events seamlessly from afar, the reality is that decisions are often being forced downwards rather than upwards. As was explored in detail in Carrick *et al.* (2009) there is often no time to wait for instructions, so the commander's intent must be interpreted by junior personnel according to the changing situation on the ground:

> The inescapable lesson of Somalia and of other recent operations, whether humanitarian assistance, peace-keeping, or traditional war fighting, is that their outcome may hinge on decisions made by small unit leaders, and by actions taken at the lowest level.
>
> (Krulak 1999)

In an age of ubiquitous media, the pressure to 'get it right' is even more profound than ever with everyone involved, from commander to private soldier, having the potential to have strategic impact with every tactical-level action. However, ethics education, as opposed to values and standards training (focusing upon right and wrong answers in specific situations), tends to be focused upon officers. They are supposed to be the ones in command and control and therefore, as issues present themselves, it makes sense to concentrate resources and time on developing the effective ethical decision-making skills of those who: (i) are likely to be in the position to make a difference, and (ii) already likely to have the combination of ability and analytical skills, in part due to their education attainment.[10] But how is one to ensure that such necessary skills are also being taught at the lower levels of rank and amongst enlisted personnel who do not carry a commission? One is immediately faced with the obvious point that the educational attainment and aptitude of many are not going to be necessarily suited to philosophical study. For example, a 2013 report warned that almost two-fifths (39 per cent) of new British Army recruits had the reading ability of an 11-year-old or lower.[11] While this might pose particular pedagogic challenges, such figures should not mean that it is impossible to do more than train in the basics – ethical decision-making can be taught in different ways to different target audiences. For example, there is no need to use Greek or Latin terminology to describe what are really very straightforward concepts in moral thinking, such as thinking through to the consequences of one's actions or being aware of things that must simply never be done, regardless of the circumstances. It is also important to note that there is a huge range of personnel, abilities, ranks and responsibilities between the least well-qualified member of the British Army and its junior officers. What can and should be delivered across that range needs to be tailored so that it is suitable to both the individuals on the receiving end and the type of ethical challenges they are likely to be faced with due to their role and likely responsibility.

Some of the challenges are certainly not unique to the UK, including the need to take into account the diverse backgrounds of those being taught, whether that is in terms of their education before joining the military, training and education within the military, or the huge potential range of operational experience that different people will have had. At the same time, it must be recognised that a great deal of it – perhaps, until very recently, the majority of that education and training – has been implicit rather than explicit, traditionally being about ethos, history, traditions, and what it is to be a 'good officer', taught by example as much as any kind of formal instruction.[12] The traditionally short commissioning process of under a year historically favoured in the UK (as opposed to the three or four year approach taken by nearly all of the UK's principal allies and partners) also means that, arguably, more emphasis needs to be placed on considering the development of individuals throughout their careers rather than expecting it to have been completed by the time the person leaves phase one training, and this is starting to be recognised now. At each stage of through-career development, it is necessary to provide an overview of the whole subject to ensure that anyone who has missed an element of it can catch up – a particular issue when courses are changing and adapting, meaning that people will inevitably miss out sometimes. But it is also important to build on what has gone before so that progression is built in – whether that is introducing new concepts and tools to aid effective ethical decision-making, or simply demonstrating how the hopefully already familiar tools can best be interpreted and applied in the context of the new environmental challenges that rank or the experience that has opened up for the individual involved.

Various reviews have been undertaken or are being undertaken at the time of writing to see how through-life training, education and development in many different areas of the British military can be supported more effectively.[13] For example, following on from work begun in 2007, the Director General of the UK Defence Academy set up a Tri-Service Moral Leadership Working Group, supported by the Military Chaplaincy, academics from King's College London and Cranfield University, and a Staff Officer from the RAF's Generic Education and Training Centre. This group managed the research and shaped a curriculum for strengthening the moral competencies appropriate to junior and senior non-commissioned Officers. This curriculum is based on the internalisation of 'Core Values' combined with developing ethical thinking, decision-making and behaviour using material specifically selected to be of direct relevance to the type of challenges faced by each cohort. While the numbers of graduates from such courses is currently severely limited by the capacity of the Armed Forces Chaplaincy Centre (who have been instrumental in setting up the courses), there are aspirations to develop the course into a 'Train the Trainer' model in the future, greatly increasing the impact across the different services in terms of numbers of people directly engaged. In a parallel development, the British Army has recently appointed its first Staff Officer grade 1 with particular responsibility for understanding and promoting ethics through the organisation, while the RAF have been focusing specifically on the coherence of through-life training and education for their personnel.

These are some of the examples of the type of investment currently being made in the area of military ethics in the UK, and meeting the challenge of producing a joined up and coherent approach to through-life training and education in ethical competency.

The current focus on developing effective ethical decision-making and the recognition that this must not be purely focused upon officers is, of course, to be warmly welcomed. But there are other changes happening that mean that such changes may prove less effective than one would hope, due to the profound changes that are taking place in the structure of the British Armed Forces. The 2010 Strategic Defence and Security Review (SDSR), entitled *Securing Britain in an Age of Uncertainty* set out a framework for providing Britain's defence with significantly fewer people in the future.[14] Taking the Army figures – the largest of the three services in personnel numbers – as of 1 April 2013, the British Army employed 104,760 full-time soldiers and 24,690 territorial soldiers (the Army Reserve).[15] The SDSR outlined a reduction in regular Army personnel to 95,000 by 2015, but in July 2011 the intention was announced to cut the trained strength even further, down to 82,000 by 2020. Clearly this is a scalar change, but in addition to the cuts in overall numbers (and the other services will also see their overall numbers of personnel fall), the structure of the military is also to change as the ratio between reservists and regulars is transformed. The trained strength of the Territorial Army is to be increased to 30,000 personnel, bringing the ratio of full-time to part-time personnel more in line with both the US and Canada.[16] The aspiration is to change the way that reservists are currently used so as to ensure that they are fully integrated and therefore better prepared for deployment alongside regular personnel.[17]

From the perspective of delivering effective PMEE, these changes pose some profound challenges, not least of which is how to provide the appropriate level of education for this growing proportion of part-time personnel. Frank Ledwidge refers to his own experience on the UK's Advanced Command and Staff Reserve Officer Course and laments the lack of critical engagement in military ethics and international law that was evident in it (Ledwidge 2011: 243). While that is an area that has certainly improved since 2006, not least in the way academics are involved with the officers to promote the appropriate level of analytical thinking, it is still the case that compared to the 'normal' advanced course for regular officers who spend nearly a year studying at the rank of Lt Colonel or equivalent, there is far less focus on the ethics and law of conflict for reservists who attend the college. However, this is more a reflection of the fact that the dedicated reservists' course, organised as it is so that people can maintain their regular employment, is therefore only about 10 per cent of the length of the regular course. The actual percentage of course time devoted to normative issues is probably similar overall, but that doesn't change the fact that there is very little time in absolute terms to engage in any subject in real depth, and that includes the ethical and legal perspectives. While this is an example taken from relatively high up the pyramid of the military rank structure, the same issues are likely to be found for reservists at all levels unless developing ethical competency is prioritised and included at the expense of other key military skills – something that is not going to be easy (or possibly even desirable) to do. Looking to allies who already have higher ratios of reservists can perhaps provide some lessons for how to do this better in the UK.[18]

At the same time, as more part-time personnel are being incorporated into military structures, an increasing number of people who turn policy into actuality in theatre are not even in the employment of the military at all but are instead employed by Private Military and Security Companies (PMSCs) or other private companies providing logistical or specialist support. This is recognised in the Total Force Concept that frames UK policy, but the actual implications of 40 per cent or more of the personnel in a theatre of operations not being in the employment of the state is a major change in operating practices and assumptions, both for the UK and her allies. The Gansler Report into US Army contracting noted back in 2007 that 'Notwithstanding there being almost as many contractor personnel in the Afghanistan/Iraq/Kuwait Theatres as there are US Military, the Operational Army does not yet recognise the impact of contracting and contractors in expeditionary operations and on mission success' (The Gansler Report 2007: 2). This raises interesting, and in some cases worrying, questions about those who are not in the direct employment of the state and the type of provision that is made for their training and education in ethics. Of course, the vast majority of PMSC employees have some kind of military background before they go into private employment as the skill set required for this type of work is difficult to acquire unless one has worked in a state's professional military. This means that, at least in theory, they will have received some training and education in military ethics and law before they left. As long as the standard of training and education received was good enough and it has been absorbed and internalised effectively, this can be considered a win-win situation from the employing state buying the service *and* the PMSC itself who effectively gets a subsidised skill set. This does, though, also highlight one of the problems in the nature of PMSC structures with regards to investing in their own people – as these are private companies in many cases employing skilled people on short-term contracts in response to a specific demand (it would be inefficient to employ them when there is no work, for example), there does not appear to be that much incentive to build and develop skill sets further in those personnel. Either they can do the job – in which case employ them – or they can't, in which case employ someone else or accept that the skill is not a priority. In such an environment, any kind of investment in individuals (other than for core team leaders who may be on longer term contracts) can too easily be seen as simply providing a benefit for the next employer.

There are a number of attempts, both by the international community and the PMSCs themselves, to change this attitude and a clear awareness, due to events such as the Black Water scandal in 2007 where a large number of Iraqi civilians were left dead, that there is huge potential for the activities of PMSCs to undermine mission success if their personnel act contrary to acceptable ethical standards.[19] States want to know that the personnel they employ, directly or indirectly, will act appropriately, and the PMSCs have a vested interest in ensuring that they do not lose future lucrative contracts. In this light, the International Code of Conduct for Private Security Providers (ICoC) is a multi-stakeholder initiative convened by the Swiss government in order to provide a framework of principles and standards based on international human rights and humanitarian law, as well as independent oversight to improve accountability.[20] Whether this will simply be seen as a 'rubber stamp' to accredit organisations who can demonstrate their paperwork is in order, or whether it can genuinely provide meaningful oversight in the area of ethical conduct has yet to be seen, but the multi-stakeholder approach does offer a way forward for thinking about how both parties – state employer and PMSC provider – might benefit from thinking about PMEE training and education in a more joined-up way.

Changes to PMEE delivery

Distance learning is a relatively new area for the UK military, but one which is attracting increased attention and resources. Some of the UK's allies, such as Canada or Australia, faced with personnel distributed over massive geographic distances, have already gained much experience of this area of pedagogy. Experience for the online Master of Arts degrees offered by King's College London, such as War in the Modern World, demonstrate that online courses are most effective when they play to the medium's strengths rather than simply trying to replicate a traditional classroom 'at a distance'.[21] So, for example, integrating media, facsimiles of original documental, video footage etc. in a way that would be less usual in a classroom can be done with very effective results using web-based technology. Podcasts can be used to provide additional information by lecturers, while 'chat forums' allow people to interact and develop ideas over a set amount of time that can then be fitted around other work commitments (suitably moderated by an academic who provides oversight, facilitation and guidance). There have been some problems with attitudes to distance learning and it is essential to be clear what the actual purpose of each course actually is. For example, Johnson-Freese notes that following the Goldwater–Nichols reforms in the US in the 1980s, distance learning became the only way that large numbers of busy officers could gain the necessary qualifications for promotion and joint positions. The result was 'a race to the bottom in terms of which service could develop the non-resident JPME [Joint Professional Military Education] programme that students could complete most quickly and with the least amount of effort' (Johnson-Freese 2013: 111). Despite this problematic experience of aspects of distance learning, the US in particular has still taken the lead when it comes to PMEE, with both Army and Navy developing substantive interactive video-based packages that can be surprisingly effective. Using a number of linked complex scenarios, actors (or rather military personnel playing specific roles) and a point of view approach, these can be pitched at different levels of rank and responsibility and are recorded in such a way that the student responses to the different situations change outcomes – perhaps not immediately, but over time. Students can see how a decision to intervene (or not) influences unit cohesion or shapes attitudes over time. The best of these packages are developed with a huge amount of input from education psychologists, military personnel and ethicists working together. Clearly, the scenarios need to be believable and draw upon situations that will be seen as credible by the students if they are to have a genuinely useful role. While accepting the limitations and pitfalls of testing in such an area, and also remembering the important distinction between allowing formative training and education that is not assessed alongside summative that is, the other advantage of such an approach is that it can provide the military with a record of an individual's moral reasoning and also their progress and development in this area over time.

While such tools should not be seen as a straightforward substitute and should still be used in conjunction with other PMEE vehicles, such new mediums for delivery, if invested in and developed appropriately, may help with providing reservists with a resource that can be accessed around regular work commitments. There is no obvious reason why such packages could not also be offered to the personnel employed by PMSCs, and could be integrated into the oversight mechanisms being developed by ICoC and others: the additional marginal cost of making the facility available to others would surely be outweighed by the benefits.

Deanna Messervey, a Canadian defence scientist working in the area of ethical climate research, argues convincingly that it is vital to apply the 'train as you intend to fight' principle to the psychological risk factors that may lead people to act unethically. She proposes as a result the following construct as an example of best practice in PMEE relating to this specific area:

First, ethics training can be conducted in a non-stressful environment so that key lessons can be absorbed (such as the impact that crowds can have on ethical decision making). This information can be repeated to increase retention of key lessons. Next, ethics training can simulate stressful situations (such as surprise and shock) to teach soldiers how to respond when confronted with ethical dilemmas under stressful conditions. This can also allow soldiers to practise coping with strong emotions such as anger. Finally, when conducting scenario-based training, soldiers and leaders can practice intervening during a staged ethical misconduct.

(Messervey and Peach 2015)

Integrating this type of approach with new mediums of delivery such as distance learning will again be challenging but, if done carefully, could also be very effective and be part of the desired end state of ensuring that PMEE can be accessible to all who require it.

As the demand for more involved ethical decision-making education within and beyond the military increases, the demand for new people to deliver it also obviously increases. Who should do this is an on-going question in PMEE (e.g. chaplains, commanding officers, lawyers, philosophers, peers etc.), but it seems obvious that non-specialists are often going to end up being involved to some extent. Where non-specialists are expected to fulfil this role, there is a need to support them appropriately with the skills and tools to answer questions, shape debates and ensure that appropriate learning outcomes are achieved. Ethics is obviously an area where a bad lead can result in very unfortunate results, so ensuring that, for example, a train-the-trainer course is sufficiently rigorous to provide enough support and that people know where to go to ask further questions, is essential. This is not an issue specific to those who may not have a pedagogic background either – even experienced academic colleagues who are comfortable with giving a lecture on one of the many areas that touch on military ethics can feel apprehensive about the huge range of possible questions that can arise as a result of engaging in ethical questioning. This can tend towards a desire to simply present the information as a straightforward brief or lecture with no opportunity to challenge or ask questions, and yet student feedback and validation, as well as academic research points towards an interactive environment being the most conducive to absorbing and internalising the key issues. People need to be able to discuss, think through and reflect on the issues at stake if the best outcome is to be achieved from the learning opportunity. This demonstrates that increasing ethical competency also entails building appropriate capacity to deliver PMEE.

Other issues

It is understandable that the military would wish to use routine and pre-deployment training to prepare its personnel for the types of challenges that they are likely to be faced with on operations. Unfortunately, this is not always as straightforward as it at first appears. The current fear of 'bad press' can actually lead to a strange disincentive to act in the right way in certain situations. One example is in the area of child soldiers. There is no doubt that this is a highly emotive area, but it also a fact that the type of conflicts that the UK and other Western states find themselves in today have a high chance of involving children carrying firearms and posing a genuine risk to the life of service personnel and civilians alike. Knowledge of disarmament, reintegration and resettlement plans and options that may be available can, in some circumstances, present positive options for transforming situations. There are sometimes alternative, less 'kinetic' options that may be available to soldiers, such as using non-lethal weapons, 'flashbangs' to disorientate and scatter or targeting of the adult leaders in the group to remove leadership and therefore threat,[22] but it would be dishonest to suggest that the dilemma of what do to when threatened with lethal force by a child is not a very real issue. How is one to prepare for such eventualities? It would appear likely that the natural response would be to pause in the face of such a threat. For obvious reasons, the costs of inaction in many security situations can be very high. On the other hand, the psychological costs of acting, even in an entirely legal and ethically appropriate way, may also be very high. A soldier in such a position must be able to draw on their training and also their moral and psychological training and support so that they can make sense of what they may be required to do to save their own life, that of their comrades or perhaps a group of other, unarmed and threatened children.

It is obvious that military personnel need to be prepared in the best way possible. Ensuring that personnel have had a chance to reflect on the situation before it occurs is a vital part of mental and physical preparedness. However, the military can be left in a situation where it feels as if it is damned if it does, and damned if it doesn't. There is a potential but completely understandable fear of generating a tabloid headline that reads 'British Soldiers Trained to Kill Kids', but a fear of bad publicity should never be an excuse to ignore an issue. Indeed, it would appear nothing short of negligent to send somebody on an operation where there was a very real risk of being faced with child soldiers and not prepare them psychologically for such an eventuality. If such training ever became an issue in the press, it must surely be dealt with more successfully by proactively engaging with the story and presenting an explanation of the training as a robust defence of best practice, rather than attempting to avoid a difficult issue by pretending it doesn't exist.

One other issue that must be considered is the perception of those being taught military ethics. Specifically, about what they are being taught, and why they are being taught it. While there is much more emphasis on PMEE now, it is also easy to create the impression that this is so that the military institution can 'cover' itself somehow. When ethical failures occur, it can then be blamed on the individual rather than anything systemic at fault as the 'box has been ticked'. Given that at least some of the current emphasis is in response to clear ethical failures in Iraq, such as the Baha Mousa case, it is easy to see how those being taught newly introduced modules may be cynical as to the motivation behind it, rather than feeling they are genuinely being equipped with the appropriate tools to deal with the types of ethical challenges they are likely to be faced with.[23] Therefore, ensuring that military personnel, reservists and members of the PMSC community see ethical education as a *useful* thing rather than as a hoop to jump through or box ticking exercise is very important, and this is only going to happen if the quality and effectiveness of that education is actually up to the appropriate standard.

Conclusions

This chapter has tried to briefly explore the changing landscape of PMEE. It is clear that the contemporary operating environment is creating new challenges, meaning that there are changes in what needs to be taught; the degree to which this is simply a matter of determining new ways to interpret and apply old, familiar principles, or whether totally new approaches need to be developed, is an ongoing debate. PMEE needs to continue while the debate rages on while ensuring that it continues to be informed by current thinking and remain as appropriate and relevant as possible to the practitioners who rely upon it. The character of those practitioners is also changing by way of the addition of increased numbers of reservists who are expected to be fully integrated with a deployed military but who may have had far less opportunity to participate in substantive training and education programmes. The growing numbers of PMSC personnel also raise issues that cannot be ignored. Ethical competency is just as important for those employed by private companies where they are involved with delivering policy on behalf of the state. Given that the numbers of contractors as a proportion of any future military deployment are likely to be at least as substantial as in the Iraq and Afghan theatres, adequate and appropriate ethics training and education needs to be considered for these personnel as well. How exactly it should be provided and who should pay for it are just some of the questions that need to be considered as a matter of urgency. At the same time, new methods of PMEE instruction are becoming possible and may provide opportunities for making the subject accessible to people who would otherwise have difficulty participating in traditional programmes. If these can be done well, avoiding the box ticking mentality of some old distance-learning programmes in other areas of military training, such methods can provide valuable new gadgets for the education toolbox.

Thankfully, there are a growing number of resources available internationally to support PMEE in light of the changing factors outlined above. At least in part thanks to this series of books, there is also a growing international network of people directly engaged in discussing both military ethics and military ethics pedagogy – how to ensure that those who need these skills are taught them in the best possible way. For example, the International Society for Military Ethics and its European Chapter are soon to be joined by Asia Pacific and South American Chapters, providing permanent forums for networks of academic, policymakers and practitioners to meet, discuss and share best practice in this area.[24] This active community will continue to facilitate interest in and focus attention on this vital policy area around the world.[25]

Notes

1 One must also mention the ongoing work of the International Society for Military Ethics (formally known as JSCOPE) and its European Chapter – Euro ISME.
2 There is debate about the degree to which the military can call itself a profession, but I agree with Stephen Coleman that at least as far as officers and senior non-commissioned officers are concerned, they have good reasons to consider that this term should apply in many different states around the world. See Coleman 2012, 36.
3 Some of these issues are explored in Whetham (2010).
4 This commitment was first codified in the 1906 and 1929 Geneva Conventions, and then restated in the 1949 Conventions and their Additional Protocols, as well as various Hague Protocols. See 1906 Geneva Convention, Article 26; 1929 Geneva Convention, Article 27; First Geneva Convention, Article 47; Second Geneva Convention, Article 48; Third Geneva Convention, Article 127; Fourth Geneva Convention, Article 144; Additional Protocol I, Article 83 (adopted by consensus); Additional Protocol II, Article 19 (adopted by consensus); Hague Convention for the Protection of Cultural Property, Article 25; Second Protocol to the Hague Convention on the Protection of Cultural Property, Article 30; Convention on Certain Conventional Weapons, Article 6.
5 See the Report of the International Commission on Intervention and State Sovereignty, http://responsibilitytoprotect.org/ICISS%20Report.pdf.
6 http://transcripts.cnn.com/TRANSCRIPTS/0209/08/le.00.html.
7 Some of these issues are addressed in A. Ellner, P. Robinson and D. Whetham (eds.) (2014) *When Soldiers Say No: Selective Conscientious Objection in the Modern Military*, Farnham, Surrey: Ashgate.
8 See George R. Lucas Jr, 'Foreword: This is Not Your Father's War' in Carrick, Connelly and Robinson, 2009, xvi.
9 See D. Whetham (2009) 'The Moral, Legal and Ethical Dimensions of War at the Joint Services Command and Staff College', in Carrick, Connelly and Robinson.
10 While not all officers are graduates, most are. They certainly also have a much greater investment of time in their military training and education than enlisted personnel.
11 Sellgren, K. (2013). 'Almost 40% of army recruits have reading age or 11, MPs warn'. *BBC News*, 10 July, online at: www.bbc.co.uk/news/education-23346693.
12 See P. Robinson, 'Ethics Training and Development in the Military', in *Parameters*, Spring 2007.
13 Some of those reviews have come in response to profound failures in this area. E.g. Aitkin, R. (2008) *The Aitken Report: An Investigation into Cases of Deliberate Abuse and Unlawful Killing in Iraq 2003 and 2004*, London: UK Ministry of Defence.
14 Securing Britain in an Age of Uncertainty: The Strategic Defence and Security Review, online at 2010 www.direct.gov.uk/prod_consum_dg/groups/dg_digital assets/@dg/@en/documents/digitalasset/dg_191634.pdf.
15 www.gov.uk/government/statistics/uk-armed-forces-annual-personnel-report-2.
16 Ministry of Defence. (2011) *The Independent Commission to Review the United Kingdom's Reserve Forces*, online at: www.gov.uk/government/uploads/system/uploads/attachment_data/file/28394/futurereserves_2020.pdf.
17 For example, see *Rifles reservists train with Regular Army colleagues*, 6 Nov 2012, online at MOD www.gov.uk/government/news/rifles-reservists-train-with-regular-army-colleagues.
18 For a good discussion of some of the issues relating to military education for reservists, see Mark Zelcer, 'Ethics for the Weekends: The Case of Reservists', in *Journal of Military Ethics*, 11(4): 333–352.
19 BBC News. 2009. 'Profile: Blackwater Worldwide', BBC News. 20 August, online at: http://news.bbc.co.uk/1/hi/7000645.stm.
20 See www.icoca.ch/en.

21 King's College London. (n.d.) 'War in a Modern World', online at: www.kcl.ac.uk/
    sspp/departments/warstudies/study/wsonline/programmes/wimw.aspx.
22 See HFM-159/RTO Task Group, *Child Soldiers as the Opposing Force*, NATO
    Research and Technology Organisation, January 2009, 3–7.
23 See Mackinlay, 2007. 'Perceptions and Misperceptions: How are International and
    UK Law Perceived to Affect Military Commanders and Their Subordinates on Opera-
    tions', *Defence Studies* 7(1): 111–160.
24 See for example, www.euroisme.eu.
25 Indeed, APAC ISME will meet for the first time in Canberra in 2017. www.apacisme.
    org/.

Bibliography

Bellamy, A. (2006) *Just Wars: From Cicero to Iraq*, Cambridge: Polity Press.
Carrick, D., Connelly, J. and Robinson, P. (eds.) (2009) *Ethics Education for Irregular
    Warfare*, Aldershot: Ashgate.
Clausewitz, C. (2007) *On War*, Oxford: Oxford University Press.
Coleman, S. (2002) *Military Ethics*, Oxford: Oxford University Press.
Coleman, S. (2012) *Military Ethics: An Introduction with Case Studies*, London: Oxford
    University Press.
Ellner, A., Robinson, P. and Whetham, D. (eds.) (2014) *When Soldiers Say No: Selective
    Conscientious Objection in the Modern Military*, Farnham, Surrey: Ashgate.
Department of the Navy (1990) US Marine Corps Small Wars Manual, Unites States
    Government Printing Office, 1940 (republished).
Grossman, D. (2007) *On Combat: The Psychology and Physiology of Deadly Conflict in
    War and Peace*, 2nd edition, Belleville, IL: PPCT Research Publications.
HFM-159/RTO Task Group. (2009) *Child Soldiers as the Opposing Force*, NATO
    Research and Technology Organisation, January.
Holliday, I. (2002) 'When is a Cause Just?', *Review of International Studies*, 28.
Ivy, G., Sudom, K., Dean, W. and Tremblay, M. (eds.) (2014) *The Human Dimensions of
    Operations: A Personnel Research Perspective*, Kingston, Ontario: Canadian Defence
    Academy Press.
Johnson-Freese, J. (2013) *Educating America's Military*, London: Routledge.
Krulak, C. (1999) 'The Strategic Corporal: Leadership in the Three Block War', Marines
    Gazette, January.
Ledwidge, F. (2011) *Losing Small Wars: British Military Failure in Iraq and Afghani-
    stan*, Padstow, Cornwall: Yale University Press.
Mackinlay, W.G.L. (2007) 'Perceptions and Misperceptions: How are International and
    UK Law Perceived to Affect Military Commanders and Their Subordinates on Opera-
    tions', *Defence Studies*, 7(1): 111–160.
Messervey, D. and Peach, J. (2015) 'Battlefield Ethics What Influences Ethical Behaviour
    on Operations', in K. Sudom, M.A. Tremblay, G. Ivey and W. Dean (eds.). *The Human
    Dimensions of Operations: A Personnel Research Perspective*, Kingston, ON: Cana-
    dian Defence Academy Press.
Ministry of Defence (2012) *Rifles Reservists Train with Regular Army Colleagues*, 6
    November, online at www.gov.uk/government/news/rifles-reservists-train-with-regular-
    army-colleagues.
The Gansler Report (2007) *Urgent Reform Required: US Army Expeditionary Contract-
    ing (Gansler) Report*, 31 October.

US Marine Corps (1990) *Small Wars Manual*, Washington DC: Department of the Navy.
Von Clausewitz. (1989) *On War*, Michael Howard and Peter Paret (ed. and trans.), Princeton: Princeton University Press.
Whetham, D. (2007) 'Killing Within the Rules', *Small Wars and Insurgencies*, 18(4), 721–733.
Whetham, D. (2009) *Just Wars and Moral Victories: Surprise, Deception and the Normative Framework of European War in the Later Middle Ages*, Leiden: Brill.
Whetham, D. (ed.) (2010) *Ethics, Law and Military Operations*, Basingstoke: Palgrave Macmillan.

Index

Page numbers in **bold** refer to figures and those with n attached refer to notes.